Modern Particle Physics

Modern Particle Physics

Alivia Snow

Larsen & Keller
www.larsen-keller.com

Modern Particle Physics
Alivia Snow
ISBN: 978-1-64172-685-6 (Hardback)

Larsen & Keller
Published by Larsen and Keller Education,
5 Penn Plaza,
19th Floor,
New York, NY 10001, USA

Cataloging-in-Publication Data

Modern particle physics / Alivia Snow.
p. cm.
Includes bibliographical references and index.
ISBN 978-1-64172-685-6
1. Particles (Nuclear physics). 2. Physics. I. Snow, Alivia.
QC793.2 .M63 2022
539.72--dc23

TABLE OF CONTENTS

PREFACE

This book aims to help a broader range of students by exploring a wide variety of significant topics related to this discipline. It will help students in achieving a higher level of understanding of the subject and excel in their respective fields. This book would not have been possible without the unwavered support of my senior professors who took out the time to provide me feedback and help me with the process. I would also like to thank my family for their patience and support.

Particle physics is a branch of physics that focuses on the study of nature of those particles which constitute matter and radiation. It explores the smallest detectable particles and the fundamental interactions that are necessary to explain their behavior. It also deals with the study of atomic constituents such as electrons, protons and neutrons. Theoretical particle physics seeks to develop models and theoretical framework as well as mathematical tools for understanding present experiments and predict future experiments. There are various applications of the concepts from this field such as producing medical isotopes for research and treatment as well as developing superconductors. The concepts included in this book on particle physics are of utmost significance and bound to provide incredible insights to readers. While understanding the long-term perspectives of the topics, it makes an effort in highlighting their impact as a modern tool for the growth of the discipline. It will provide comprehensive knowledge to the readers.

A brief overview of the book contents is provided below:

Chapter – Introduction

Particle physics is the branch of physics which deals with the study of the nature of the particles of radiation and matter. Some of the areas which are studied within this discipline are antiparticles, the four forces, particle conservation laws, etc. This is an introductory chapter which will briefly introduce about particle physics.

Chapter – Elementary Particles

The subatomic particles with no sub structures are known as elementary particles. Fermions and bosons are the two primary types of elementary particles. This chapter has been carefully written to provide an easy understanding of these types and related concepts of elementary particles.

Chapter – Quantum Electrodynamics

Quantum electrodynamics is the quantum field theory of electrdynamics. It describes the interaction between light and matter. It includes Dirac equation, the Feynman diagrams and Casimir's trick. This chapter has been carefully written to provide an easy understanding of these aspects of quantum electrodynamics.

Chapter - Quantum Chromodynamics

Quantum chromodynamics (QCD) is a theory which deals with the strong interaction between glucons and quarks. QCD matter and lattice QCD are some of the methods that fall under its domain. The chapter closely examines these methods of quantum chromodynamics to provide an extensive understanding of the subject.

Chapter - Strong Interactions

Strong interaction is referred to the mechanism that causes a strong nuclear force. The strong interactions of nucleons, isospin, gluons, baryons, etc. are studied within this subject. The topics elaborated in this chapter will help in gaining a better perspective about strong interactions associated with particle physics.

Chapter - Weak Interactions

The mechanism of interaction between subatomic particles responsible for the radioactive decay of atoms is known as weak interaction. It is primarily understood in terms of electroweak theory and consists of properties like weak hypercharge and weak isospin. This chapter has been carefully written to provide an easy understanding of weak interactions.

Alivia Snow

Introduction 1

- Particle
- Antiparticle
- The Four Forces
- Particle Physics
- Unparticle Physics
- The Standard Model
- Weak Interaction
- Particle Conservation Laws

Particle physics is the branch of physics which deals with the study of the nature of the particles of radiation and matter. Some of the areas which are studied within this discipline are antiparticles, the four forces, particle conservation laws, etc. This is an introductory chapter which will briefly introduce about particle physics.

Particle

A particle refers to a quantity of matter that is used by scientists to construct theories about their field of study. There is no particular size restriction on defining a particle. Astronomers can define particles to be stars in the night sky, while physicists can define particles to be electrons. It mostly depends on the scientific field and theory under development.

Particles Properties and Classification

'If there is such a large variety of objects that can be thought of as particles, is there a common theme among all of them?' Well, there is, and we will now discuss it.

Scientists often think of particles as point-like objects, meaning that they are considered shapeless for the purposes of the theory. For example, when a chemist is studying the properties of gas particles in a container, he or she would think of them as little shapeless objects that bounce against the walls of their container. If an engineer is studying traffic flow on a busy street, he or she would consider all the vehicles to be particles, disregarding whether a particular vehicle is a bus, car, or motorcycle.

Although all of the previous examples described particles in motion, it is important to note that particles can be permanently stationary objects, at least for the purposes of the theory. For example, the carbon atoms that make up graphite, a primary constituent of pencil lead, can be thought of as particles. Atomic and subatomic particles, which are very important in many scientic fields.

Particle Classification

The four fundamental interactions or forces that govern the behavior of elementary particles are listed below:

- The strong force (It holds the nucleus together).
- The electromagnetic force (It causes interactions between charges).
- The weak force (It causes beta decay).
- The gravitational force (It causes interaction between states with energy.

A given particle may not necessarily be subject to all four interactions. Neutrinos, for example, experience only the weak and gravitational interaction.

The fundamental particles may be classified into groups in several ways. First, all particles are classified into fermions, which obey Fermi-Dirac statistics and bosons, which obey Bose-Einstein statistics. Fermions have half-integer spin, while bosons have integer spin. All the fundamental fermions have spin 1/2. Electrons and nucleons are fermions with spin 1/2. The fundamental bosons have mostly spin 1. This includes the photon. The pion has spin 0, while the graviton has spin 2. There are also three particles, the W+, W− and Z0 bosons, which are spin 1. They are the carriers of the weak interactions.

We can also classify the particles according to their interactions.

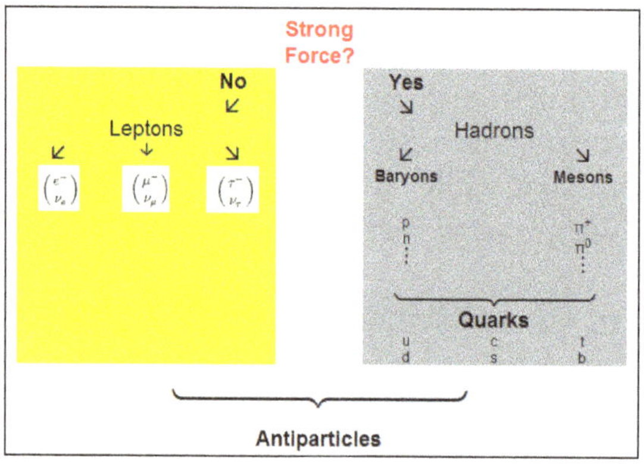

The electron and the neutrino are members of a family of leptons. Originally leptons meant "light particles", as opposed to baryons, or heavy particles, which referred initially to the proton and neutron. The pion, or pi-meson, and another particle called the muon or mu-meson, were called mesons, or medium-weight particles, because their masses, a few hundred times heavier than the electron but six times lighter than a proton, were in the middle. But that distinction turned out not to be very useful. We now recognize the muon to be almost the same as an electron, and the leptons now consist of three "generations" of pairs of particles,

$$\begin{pmatrix} e^- \\ \hline \nu_e \end{pmatrix}, \begin{pmatrix} \mu^- \\ \hline \nu_\mu \end{pmatrix}, \begin{pmatrix} T^- \\ \hline V_T \end{pmatrix},$$

with the heaviest of these, the tau lepton T-, being almost twice as massive as the proton.

The leptons are distinguished from other particles called hadrons in that leptons do not participate in strong interactions. The bottom lepton in each of the three "doublets" shown above is not only neutral, but also has a very small mass. Neutrinos had been considered massless for many years, but more recent experiments have shown their mass to be non-zero.

Particle			Associated neutrino		
Name	Charge (e)	Mass (MeV)	Name	Charge (e)	Mass (MeV)
Electron (e^-)	-1	0.511	Electron neutrino (ν_e)	0	< 0.000003
Muon (μ^-)	-1	105.6	Muon neutrino (ν_μ)	0	< 0.19
Tau (τ^-)	-1	1777	Tau neutrino (ν_τ)	0	< 18.2

Hadrons are strongly interacting particles. They are divided into baryons and mesons. The baryons are a class of fermions, including the proton and neutron, and other particles which in a decay always produce another baryon, and ultimately a proton. The mesons, are bosons. In addition to the pion, there are other spin 0 particles, four kaons and two eta mesons, and a number of spin one hadrons, including the three rho mesons, which like the pion come in charges 1 and 0. Mesons can decay without necessarily producing other hadrons.

Table of some baryons:

Particle	Symbol	Quark Content	Mass MeV/c²	Mean lifetime (s)	Decays to
Proton	p	uud	938.3	Stable	Unobserved
Neutron	n	ddu	939.6	885.7±0.8	$p + e^- + \nu_e$
Delta	Δ^{++}	uuu	1232	6×10^{-24}	$\pi^+ + p$
Delta	Δ^+	uud	1232	6×10^{-24}	$\pi^+ + n$ or $\pi^0 + p$
Delta	Δ^0	udd	1232	6×10^{-24}	$\pi^0 + n$ or $\pi^- + p$
Delta	Δ^-	ddd	1232	6×10^{-24}	$\pi^- + n$
Lambda	Λ^0	uds	1115.7	2.60×10^{-10}	$\pi^- + p$ or $\pi^0 + n$
Sigma	Σ^+	uus	1189.4	0.8×10^{-10}	$\pi^0 + p$ or $\pi^+ + n$
Sigma	Σ^0	uds	1192.5	6×10^{-20}	$\Lambda^0 + \gamma$
Sigma	Σ^-	dds	1197.4	1.5×10^{-10}	$\pi^- + n$
Xi	Ξ^0	uss	1315	2.9×10^{-10}	$\Lambda^0 + \pi^0$
Xi	Ξ^-	dss	1321	1.6×10^{-10}	$\Lambda^0 + \pi^-$
Omega	Ω^-	sss	1672	0.82×10^{-10}	$\Lambda^0 + K^-$ or $\Xi^0 + \pi^-$

Table of some Mesons:

Particle	Symbol	Anti- particle	Quark Content	Mass MeV/c^2	Mean lifetime (s)	Principal decays
Charged Pion	π^+	π^-	ud	139.6	2.60×10^{-8}	$\mu^+ + \nu_\mu$
Neutral Pion	π^0	Self	uu - dd	135.0	0.84×10^{-16}	2γ
Charged Kaon	K^+	K^-	$u\bar{s}$	493.7	1.24×10^{-8}	$\mu^+ + \nu_\mu$ or $\pi^+ + \pi_0$
Neutral Kaon	K^0	K^0	ds	497.7		
Eta	η	Self	uu + dd - 2ss	547.8	5×10^{-19}	
Eta Prime	η'	Self	uu + dd + ss	957.6	3×10^{-2}	

Each elementary particle is associated with an antiparticle with the same mass and opposite charge. Some particles, such as the photon, are identical to their antiparticle. Such particles must be neutral, but not all neutral particles are identical to their antiparticle. Particle-antiparticle pairs can annihilate each other if they are in appropriate quantum states, releasing an amount of energy equal to twice the rest energy of the particle. They can also be produced in various processes, if enough energy is available. The minimum amount of energy needed is twice the rest energy of the particle, if momentum conservation allows the particle-antiparticle pair to be produced at rest. Most often the antiparticle is denoted by the same symbol as the particle, but with a line over the symbol. For example, the antiparticle of the proton p, is denoted by p.

Protons and neutrons are made of still smaller particles called quarks. At this time it appears that the two basic constituents of matter are the leptons and the quarks. There are believed to be six types of each. Each quark type is called a flavor, there are six quark flavors. Each type of lepton and quark also has a corresponding antiparticle, a particle that has the same mass but opposite electrical charge and magnetic moment. An isolated quark has never been found, quarks appear to almost always be found in pairs or triplets with other quarks and antiquarks. The resulting particles are the hadrons, more than 200 of which have been identified. Baryons are made up of 3 quarks, and mesons are made up of a quark and an anti-quark. Baryons are fermions and mesons are bosons. Two theoretically predicted five-quark particles, called pentaquarks, have been produced in the laboratory. Four- and six-quark particles are also predicted but have not been found.

The six quarks have been named up, down, charm, strange, top, and bottom. The top quark, which has a mass greater than an entire atom of gold, is about 35 times more massive than the next biggest quark and may be the heaviest particle nature has ever created. The quarks found in ordinary matter are the up and down quarks, from which protons and neutrons are made. A proton consists of two up quarks and a down quark, and a neutron consists of two down quarks and an up quark. The pentaquark consists of two up quarks, two down quarks, and the strange antiquark. Quarks have fractional charges of one third or two thirds of the basic charge of the electron or proton. Particles made from quarks always have integer charge.

Table of Quarks:

Name	Symbol	Charge (e)	Spin	Mass MeV/c^2	Strangeness	Baryon number	Lepton number
Up	u	+2/3	1/2	1.7-3.3	0	1/3	0
Down	d	-1/3	1/2	4.1-5.8	0	1/3	0

Strange	s	-1/3	1/2	101	-1	1/3	0
Charm	c	+2/3	1/2	1270	0	1/3	0
Bottom	b	-1/3	1/2	4190-4670	0	1/3	0
Top	t	+2/3	1/2	172000	0	1/3	0

In the current theory, known as the Standard Model there are 12 fundamental matter particle types and their corresponding antiparticles. In addition, there are gluons, photons, and W and Z bosons, the force carrier particles that are responsible for strong, electromagnetic, and weak interactions respectively. These force carriers are also fundamental particles.

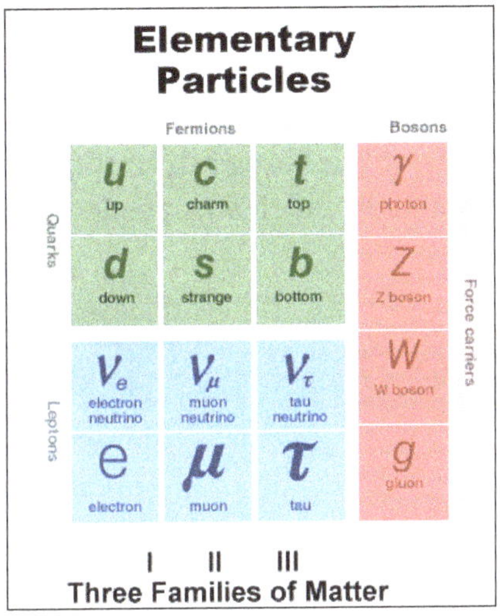

All we know is that quarks and leptons are smaller than 10-19 meters in radius. As far as we can tell, they have no internal structure or even any size. It is possible that future evidence will, once again, show our understanding to be incomplete and demonstrate that there is substructure within the particles that we now view as fundamental.

The first subatomic particle to be discovered was the electron, identified in 1897 by J. J. Thomson. After the nucleus of the atom was discovered in 1911 by Ernest Rutherford, the nucleus of ordinary hydrogen was recognized to be a single proton. In 1932 the neutron was discovered. An atom was seen to consist of a central nucleus containing protons and neutrons, surrounded by orbiting electrons. However, other elementary particles not found in ordinary atoms immediately began to appear.

In 1928 the relativistic quantum theory of P. A. M. Dirac predicted the existence of a positively charged electron, or positron, which is the antiparticle of the electron. It was first detected in 1932. Difficulties in explaining beta decay led to the prediction of the neutrino in 1930, and by 1934 the existence of the neutrino was firmly established in theory, although it was not actually detected until 1956. Another particle was also added to the list, the photon, which had been first suggested by Einstein in 1905 as part of his quantum theory of the photoelectric effect.

The next particles discovered were related to attempts to explain the strong interactions, or strong nuclear force, binding nucleons together in an atomic nucleus. In 1935 Hideki Yukawa suggested that a meson, a charged particle with a mass intermediate between those of the electron and the proton, might be exchanged between nucleons. The meson emitted by one nucleon would be absorbed by another nucleon. This would produce a strong force between the nucleons, analogous to the force produced by the exchange of photons between charged particles interacting through the electromagnetic force. It is now known that the strong force is mediated by the gluon. The following year a particle of approximately the required mass, about 200 times that of the electron, was discovered and named the mu-meson, or muon. However, its behavior did not conform to that of the theoretical particle. In 1947 the particle predicted by Yukawa was finally discovered and named the pi-meson, or pion. Both the muon and the pion were first observed in cosmic rays. Further studies of cosmic rays turned up more particles. By the 1950s these elementary particles were also being observed in the laboratory as a result of particle collisions in particle accelerators.

By the early 1960s over 30 "fundamental particles" had been found. A rigorous way of classifying them was needed. Were there any symmetries or patterns? Murray Gell-Mann believed that a framework for such patterns could be found in the mathematical structure of groups. A symmetry group called SU(3) offered patterns he was looking for. In 1961, after grouping the known particles, he predicted the existence of the η particle which was needed to complete a pattern. The η particle was discovered a few months later.

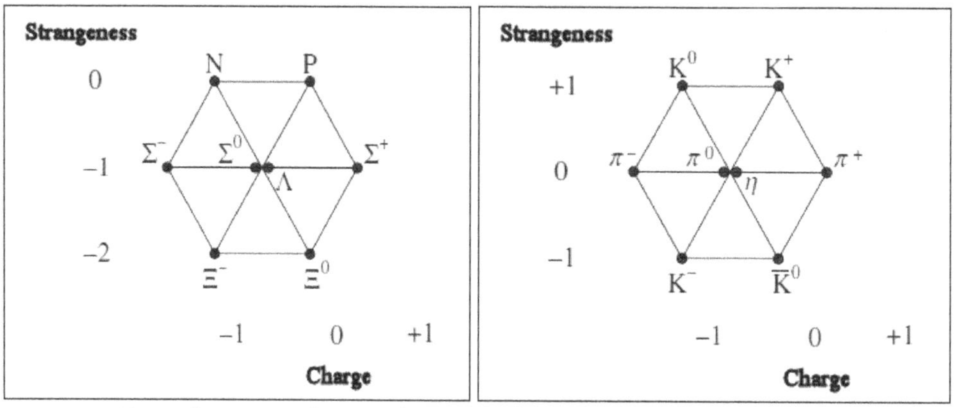

Example patterns for some baryons and mesons (The Eightfold Way).

After finding the patterns an explanation was needed. In 1964 Gell-Mann published a short article showing that the patterns could be produced if the known particles were viewed as combinations of 3 fundamental subunits with fractional charge, the up, down, and strange quarks and their anti-quarks. There were however problems with the Pauli Exclusion Principle. The quarks are spin 1/2 fermions and the Δ++(uuu) and the Ω-(sss) seemed to contain at least two quarks with exactly the same quantum numbers. The quark theory was not really accepted until deep inelastic scattering experiments revealed structure inside the protons in the later 1960s.

The charm quark was discovered in 1974, the bottom quark in 1977, and the top quark in 1995. The tau particle was detected in a series of experiments between 1974 and 1977 and the discovery of the tau neutrino was announced in 2000. It was the last of the particles in the Standard Model of elementary particles to be detected.

One of the current frontiers in the study of elementary particles concerns the interface between that discipline and cosmology. The known quarks and leptons, for instance, are typically grouped in three families, where each family contains two quarks and two leptons. Investigators have wondered whether additional families of elementary particles might be found. Recent work in cosmology pertaining to the evolution of the universe has suggested that there could be no more families than four, and the cosmological theory has been substantiated by experimental work at the Stanford Linear Accelerator and at CERN, which indicates that there are no families of elementary particles other than the three that are known today. For example, detailed studies of Zo decays at CERN revealed that there can be no more than three different kinds of neutrino. If there was a fourth, or a fifth, further decay routes would be open to the Zo which would affect its measured lifetime.

Antiparticle

In particle physics, every type of particle has an associated antiparticle with the same mass but with opposite physical charges (such as electric charge). For example, the antiparticle of the electron is the antielectron (which is often referred to as *positron*). While the electron has a negative electric charge, the positron has a positive electric charge, and is produced naturally in certain types of radioactive decay. The opposite is also true: the antiparticle of the positron is the electron.

Some particles, such as the photon, are their own antiparticle. Otherwise, for each pair of antiparticle partners, one is designated as normal matter (the kind all matter usually interacted with is made of), and the other as *antimatter*.

Particle–antiparticle pairs can annihilate each other, producing photons; since the charges of the particle and antiparticle are opposite, total charge is conserved. For example, the positrons produced in natural radioactive decay quickly annihilate themselves with electrons, producing pairs of gamma rays, a process exploited in positron emission tomography.

The laws of nature are very nearly symmetrical with respect to particles and antiparticles. For example, an antiproton and a positron can form an antihydrogen atom, which is believed to have the same properties as a hydrogen atom. This leads to the question of why the formation of matter after the Big Bang resulted in a universe consisting almost entirely of matter, rather than being a half-and-half mixture of matter and antimatter. The discovery of Charge Parity violation helped to shed light on this problem by showing that this symmetry, originally thought to be perfect, was only approximate.

Because charge is conserved, it is not possible to create an antiparticle without either destroying another particle of the same charge (as is for instance the case when antiparticles are produced naturally via beta decay or the collision of cosmic rays with Earth's atmosphere), or by the simultaneous creation of both a particle *and* its antiparticle, which can occur in particle accelerators such as the Large Hadron Collider at CERN.

Although particles and their antiparticles have opposite charges, electrically neutral particles need not be identical to their antiparticles. The neutron, for example, is made out of quarks, the antineutron from antiquarks, and they are distinguishable from one another because neutrons and antineutrons annihilate each other upon contact. However, other neutral particles are their own antiparticles, such as photons, Z^o bosons, π^omesons, and hypothetical gravitons and some hypothetical WIMPs.

Experiment

In 1932, soon after the prediction of positrons by Paul Dirac, Carl D. Anderson found that cosmic-ray collisions produced these particles in a cloud chamber— a particle detector in which moving electrons (or positrons) leave behind trails as they move through the gas. The electric charge-to-mass ratio of a particle can be measured by observing the radius of curling of its cloud-chamber track in a magnetic field. Positrons, because of the direction that their paths curled, were at first mistaken for electrons travelling in the opposite direction. Positron paths in a cloud-chamber trace the same helical path as an electron but rotate in the opposite direction with respect to the magnetic field direction due to their having the same magnitude of charge-to-mass ratio but with opposite charge and, therefore, opposite signed charge-to-mass ratios.

The antiproton and antineutron were found by Emilio Segrè and Owen Chamberlain in 1955 at the University of California, Berkeley. Since then, the antiparticles of many other subatomic particles have been created in particle accelerator experiments. In recent years, complete atoms of antimatter have been assembled out of antiprotons and positrons, collected in electromagnetic traps.

Dirac Hole Theory

Solutions of the Dirac equation contained negative energy quantum states. As a result, an electron could always radiate energy and fall into a negative energy state. Even worse, it could keep radiating infinite amounts of energy because there were infinitely many negative energy states available. To prevent this unphysical situation from happening, Dirac proposed that a "sea" of negative-energy electrons fills the universe, already occupying all of the lower-energy states so that, due to the Pauli exclusion principle, no other electron could fall into them. Sometimes, however, one of these negative-energy particles could be lifted out of this Dirac sea to become a positive-energy particle. But, when lifted out, it would leave behind a *hole* in the sea that would act exactly like a positive-energy electron with a reversed charge. These holes were interpreted as "negative-energy electrons" by Paul Dirac and by mistake he identified them with protons in his 1930 paper A Theory of Electrons and Protons However, these "negative-energy electrons" turned out to be positrons, and not protons.

This picture implied an infinite negative charge for the universe—a problem of which Dirac was aware. Dirac tried to argue that we would perceive this as the normal state of zero charge. Another difficulty was the difference in masses of the electron and the proton. Dirac tried to argue that this was due to the electromagnetic interactions with the sea, until Hermann Weyl proved that hole theory was completely symmetric between negative and positive charges. Dirac also predicted a reaction $e^- + p^+ \rightarrow \gamma + \gamma$, where an electron and a proton annihilate to give two photons. Robert Oppenheimer and Igor Tamm proved that this would cause ordinary matter to disappear too fast. A year later, in 1931, Dirac modified his theory and postulated the positron, a new particle of the same mass as the electron. The discovery of this particle the next year removed the last two objections to his theory.

Within Dirac's theory, the problem of infinite charge of the universe remains. Some bosons also have antiparticles, but since bosons do not obey the Pauli exclusion principle (only fermions do), hole theory does not work for them. A unified interpretation of antiparticles is now available in quantum field theory, which solves both these problems by describing antimatter as negative energy states of the same underlying matter field i.e. particles moving backwards in time.

Particle–antiparticle Annihilation

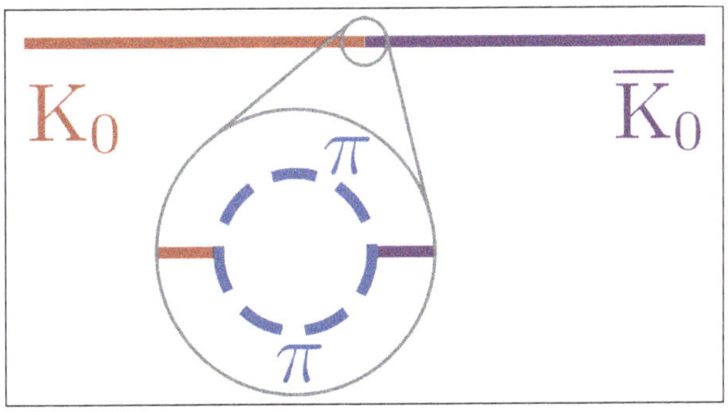

An example of a virtual pion pair that influences the propagation of a kaon, causing a neutral kaon to *mix* with the antikaon. This is an example of renormalization in quantum field theory— the field theory being necessary because of the change in particle number.

If a particle and antiparticle are in the appropriate quantum states, then they can annihilate each other and produce other particles. Reactions such as $e^- + e^+ \to \gamma + \gamma$ (the two-photon annihilation of an electron-positron pair) are an example. The single-photon annihilation of an electron-positron pair, $e^- + e^+ \to \gamma$, cannot occur in free space because it is impossible to conserve energy and momentum together in this process. However, in the Coulomb field of a nucleus the translational invariance is broken and single-photon annihilation may occur. The reverse reaction (in free space, without an atomic nucleus) is also impossible for this reason. In quantum field theory, this process is allowed only as an intermediate quantum state for times short enough that the violation of energy conservation can be accommodated by the uncertainty principle. This opens the way for virtual pair production or annihilation in which a one particle quantum state may *fluctuate* into a two particle state and back. These processes are important in the vacuum state and renormalization of a quantum field theory. It also opens the way for neutral particle mixing through processes such as the one pictured here, which is a complicated example of mass renormalization.

Properties

Quantum states of a particle and an antiparticle can be interchanged by applying the charge conjugation (C), parity (P), and time reversal (T) operators. If $|p,\sigma,n\rangle$ denotes the quantum state of a particle (n) with momentum p, spin J whose component in the z-direction is σ, then one has:

$$CPT\,|p,\sigma,n\rangle = (-1)^{J-\sigma}\,|p,-\sigma,n^c\rangle,$$

where n^c denotes the charge conjugate state, that is, the antiparticle. This behaviour under CPT symmetry is the same as the statement that the particle and its antiparticle lie in the same irreducible representation of the Poincaré group. Properties of antiparticles can be related to those of particles through this. If T is a good symmetry of the dynamics, then:

$$T\,|p,\sigma,n\rangle \propto |-p,-\sigma,n\rangle,$$

$$CP\,|\,p,\sigma,n\rangle \propto |-p,\sigma,n^c\rangle,$$

$$C\,|\,p,\sigma,n\rangle \propto |\,p,\sigma,n^c\rangle,$$

where the proportionality sign indicates that there might be a phase on the right hand side. In other words, particle and antiparticle must have:

- The same mass m,
- The same spin state j,
- Opposite electric charges q and -q.

Quantum Field Theory

One may try to quantize an electron field without mixing the annihilation and creation operators by writing,

$$\psi(x) = \sum_k u_k(x) a_k e^{-iE(k)t},$$

where we use the symbol *k* to denote the quantum numbers *p* and σ of the previous topic and the sign of the energy, *E(k)*, and a_k denotes the corresponding annihilation operators. Of course, since we are dealing with fermions, we have to have the operators satisfy canonical anti-commutation relations. However, if one now writes down the Hamiltonian $H = \sum_k E(k) a_k^\dagger a_k$, then one sees immediately that the expectation value of *H* need not be positive. This is because *E(k)* can have any sign whatsoever, and the combination of creation and annihilation operators has expectation value 1 or 0.

So one has to introduce the charge conjugate *antiparticle* field, with its own creation and annihilation operators satisfying the relations,

$$b_{k'} = a_k^\dagger \text{ and } b_{k'}^\dagger = a_k,$$

where *k* has the same *p*, and opposite σ and sign of the energy. Then one can rewrite the field in the form,

$$\psi(x) = \sum_{k_+} u_k(x) a_k e^{-iE(k)t} + \sum_{k_-} u_k(x) b_k^\dagger e^{-iE(k)t},$$

where the first sum is over positive energy states and the second over those of negative energy. The energy becomes,

$$H = \sum_{k_+} E_k a_k^\dagger a_k + \sum_{k_-} |E(k)|\, b_k^\dagger b_k + E_0,$$

where E_o is an infinite negative constant. The vacuum state is defined as the state with no particle or antiparticle *i.e.,* $a_k\,|0\rangle = 0$ and $b_k\,|0\rangle = 0$. Then the energy of the vacuum is exactly E_o. Since all energies are measured relative to the vacuum, H is positive definite. Analysis of the properties of a_k and b_k shows that one is the annihilation operator for particles and the other for antiparticles. This is the case of a fermion.

This approach is due to Vladimir Fock, Wendell Furry and Robert Oppenheimer. If one quantizes a real scalar field, then one finds that there is only one kind of annihilation operator; therefore, real scalar fields describe neutral bosons. Since complex scalar fields admit two different kinds of annihilation operators, which are related by conjugation, such fields describe charged bosons.

Feynman–Stueckelberg Interpretation

By considering the propagation of the negative energy modes of the electron field backward in time, Ernst Stueckelberg reached a pictorial understanding of the fact that the particle and antiparticle have equal mass m and spin J but opposite charges q. This allowed him to rewrite perturbation theory precisely in the form of diagrams. Richard Feynman later gave an independent systematic derivation of these diagrams from a particle formalism, and they are now called Feynman diagrams. Each line of a diagram represents a particle propagating either backward or forward in time. This technique is the most widespread method of computing amplitudes in quantum field theory today.

Since this picture was first developed by Stueckelberg, and acquired its modern form in Feynman's work, it is called the Feynman–Stueckelberg interpretation of antiparticles to honor both scientists

The Four Forces

The Four Fundamental Forces of Nature are Gravitational force, Weak Nuclear force, Electromagnetic force and Strong Nuclear force. The weak and strong forces are effective only over a very short range and dominate only at the level of subatomic particles. Gravity and Electromagnetic force have infinite range.

The four fundamental forces and their strengths:

1. Gravitational Force – Weakest force; but infinite range. (Not part of standard model)
2. Weak Nuclear Force – Next weakest; but short range.
3. Electromagnetic Force – Stronger, with infinite range.
4. Strong Nuclear Force – Strongest; but short range.

Gravitational Force

The gravitational force is weak, but very long ranged. Furthermore, it is always attractive. It acts between any two pieces of matter in the Universe since mass is its source.

Weak Nuclear Force

The weak force is responsible for radioactive decay and neutrino interactions. It has a very short range and. As its name indicates, it is very weak. The weak force causes Beta decay ie. the conversion of a neutron into a proton, an electron and an antineutrino.

Electromagnetic Force

The electromagnetic force causes electric and magnetic effects such as the repulsion between like electrical charges or the interaction of bar magnets. It is long-ranged, but much weaker than the strong force. It can be attractive or repulsive, and acts only between pieces of matter carrying electrical charge. Electricity, magnetism, and light are all produced by this force.

Strong Nuclear Force

The strong interaction is very strong, but very short-ranged. It is responsible for holding the nuclei of atoms together. It is basically attractive, but can be effectively repulsive in some circumstances. The strong force is 'carried' by particles called gluons; that is, when two particles interact through the strong force, they do so by exchanging gluons. Thus, the quarks inside of the protons and neutrons are bound together by the exchange of the strong nuclear force.

Note : While they are close together the quarks experience little force, but as they separate the force between them grows rapidly, pulling them back together. To separate two quarks completely would require far more energy than any possible particle accelerator could provide.

Fundamental Force Particles

Force	Particles Experiencing	Force Carrier Particle	Range	Relative Strength*
Gravity acts between objects with mass	all particles with mass	graviton (not yet observed)	infinity	much weaker
Weak Force governs particle decay	quarks and leptons	W^+, W^-, Z^0 (W and Z)	short range	↓
Electromagnetism acts between electrically charged particles	electrically charged	γ (photon)	infinity	↓
Strong Force** binds quarks together	quarks and gluons	g (gluon)	short range	much stronger

Electroweak Theory and Grand Unification Theories (GUT)

There is a speculation, that In the very early Universe when temperatures were very high (the Planck Scale) all four forces were unified into a single force. Then, as the temperature dropped, gravitation separated first and then the other 3 forces separated. Even then, the weak, electromagnetic, and strong forces were unified into a single force. When the temperature dropped these forces got separated from each other, with the strong force separating first and then at a still lower temperature the electromagnetic and weak forces separating to leave us with the 4 distinct forces that we see in our present Universe. The process of the forces separating from each other is called spontaneous symmetry breaking.

- The weak and electromagnetic interactions have been unified under Standard Electroweak Theory, or sometimes just the Standard Model.
- Grand unification theories attempt to treat both strong and electroweak interactions under the same mathematical structure. Unification of Weak forces and strong forces PS: Attempts to include gravitation in this picture have not yet been successful.
- Theories that add gravity to the mix and try to unify all four fundamental forces into a single force are called Superunified Theories.
- PS: Grand Unified and Superunified Theories remain theoretical speculations that are as yet unproven, but there is strong experimental evidence for the unification of the electromagnetic and weak interactions in the Standard Electroweak Theory. Furthermore, although GUTs are not proven experimentally, there is strong circumstantial evidence to suggest that a theory at least like a Grand Unified Theory is required to make sense of the Universe.

Particle Physics

Particle physics is a branch of physics that studies the elementary constituents of matter and radiation, and the interactions between them.

It is also called "high energy physics", because many elementary particles do not occur under normal circumstances in nature, but can be created and detected during energetic collisions of other particles, as is done in particle accelerators.

Modern particle physics research is focused on subatomic particles, which have less structure than atoms.

These include atomic constituents such as electrons, protons, and neutrons (protons and neutrons are actually composite particles, made up of quarks), particles produced by radiative and scattering processes, such as photons, neutrinos, and muons, as well as a wide range of exotic particles.

Strictly speaking, the term particle is a misnomer because the dynamics of particle physics are governed by quantum mechanics. As such, they exhibit wave-particle duality, displaying particle-like behavior under certain experimental conditions and wave-like behavior in others (more technically they are described by state vectors in a Hilbert space).

All the particles and their interactions observed to date can be described by a quantum field theory called the Standard Model.

The Standard Model has 40 species of elementary particles (24 fermions, 12 vector bosons, and 4 scalars), which can combine to form composite particles, accounting for the hundreds of other species of particles discovered since the 1960s.

Unparticle Physics

In theoretical physics, unparticle physics is a speculative theory that conjectures a form of matter that cannot be explained in terms of particles using the Standard Model of particle physics, because its components are scale invariant.

Howard Georgi proposed this theory in two 2007 papers, "Unparticle Physics" and "Another Odd Thing About Unparticle Physics". His papers were followed by further work by other researchers into the properties and phenomenology of unparticle physics and its potential impact on particle physics, astrophysics, cosmology, CP violation, lepton flavour violation, muon decay, neutrino oscillations, and supersymmetry.

Properties

Unparticles would have properties in common with neutrinos, which have almost zero mass and are therefore nearly scale invariant. Neutrinos barely interact with matter – most of the time physicists can infer their presence only by calculating the "missing" energy and momentum after an interaction. By looking at the same interaction many times, a probability distribution is built up that tells more specifically how many and what sort of neutrinos are involved. They couple very weakly to ordinary matter at low energies, and the effect of the coupling increases as the energy increases.

A similar technique could be used to search for evidence of unparticles. According to scale invariance, a distribution containing unparticles would become apparent because it would resemble a distribution for a fractional number of massless particles.

This scale invariant sector would interact very weakly with the rest of the Standard Model, making it possible to observe evidence for unparticle stuff, if it exists. The unparticle theory is a high-energy theory that contains both Standard Model fields and Banks–Zaks fields, which have scale-invariant behavior at an infrared point. The two fields can interact through the interactions of ordinary particles if the energy of the interaction is sufficiently high.

These particle interactions would appear to have "missing" energy and momentum that would not be detected by the experimental apparatus. Certain distinct distributions of missing energy would signify the production of unparticle stuff. If such signatures are not observed, bounds on the model can be set and refined.

Experimental Indications

Unparticle physics has been proposed as an explanation for anomalies in superconducting cuprate materials, where the charge measured by ARPES appears to exceed predictions from Luttinger's theorem for the quantity of electrons.

The Standard Model

The Standard Model of particle physics is the theory describing three of the four known fundamental forces (the electromagnetic, weak, and strong interactions, and not including the gravitational force) in the universe, as well as classifying all known elementary particles. It was developed in stages throughout the latter half of the 20th century, through the work of many scientists around the world, with the current formulation being finalized in the mid-1970s upon experimental confirmation of the existence of quarks. Since then, confirmation of the top quark, the tau neutrino, and the Higgs boson have added further credence to the Standard Model. In addition, the Standard Model has predicted various properties of weak neutral currents and the W and Z bosons with great accuracy.

Although the Standard Model is believed to be theoretically self-consistent and has demonstrated huge successes in providing experimental predictions, it leaves some phenomena unexplained and falls short of being a complete theory of fundamental interactions. It does not fully explain baryon asymmetry, incorporate the full theory of gravitation as described by general relativity, or account for the accelerating expansion of the Universe as possibly described by dark energy. The model does not contain any viable dark matter particle that possesses all of the required properties deduced from observational cosmology. It also does not incorporate neutrino oscillations and their non-zero masses.

The development of the Standard Model was driven by theoretical and experimental particle physicists alike. For theorists, the Standard Model is a paradigm of a quantum field theory, which exhibits a wide range of physics including spontaneous symmetry breaking, anomalies and non-perturbative behavior. It is used as a basis for building more exotic models that incorporate hypothetical particles, extra dimensions, and elaborate symmetries (such as supersymmetry) in an attempt to explain experimental results at variance with the Standard Model, such as the existence of dark matter and neutrino oscillations.

At present, matter and energy are best understood in terms of the kinematics and interactions of elementary particles. To date, physics has reduced the laws governing the behavior and interaction of all known forms of matter and energy to a small set of fundamental laws and theories. A major goal of physics is to find the "common ground" that would unite all of these theories into one integrated theory of everything, of which all the other known laws would be special cases, and from which the behavior of all matter and energy could be derived (at least in principle).

Particle Content

The Standard Model includes members of several classes of elementary particles, which in turn can be distinguished by other characteristics, such as color charge.

All particles can be summarized as follows:

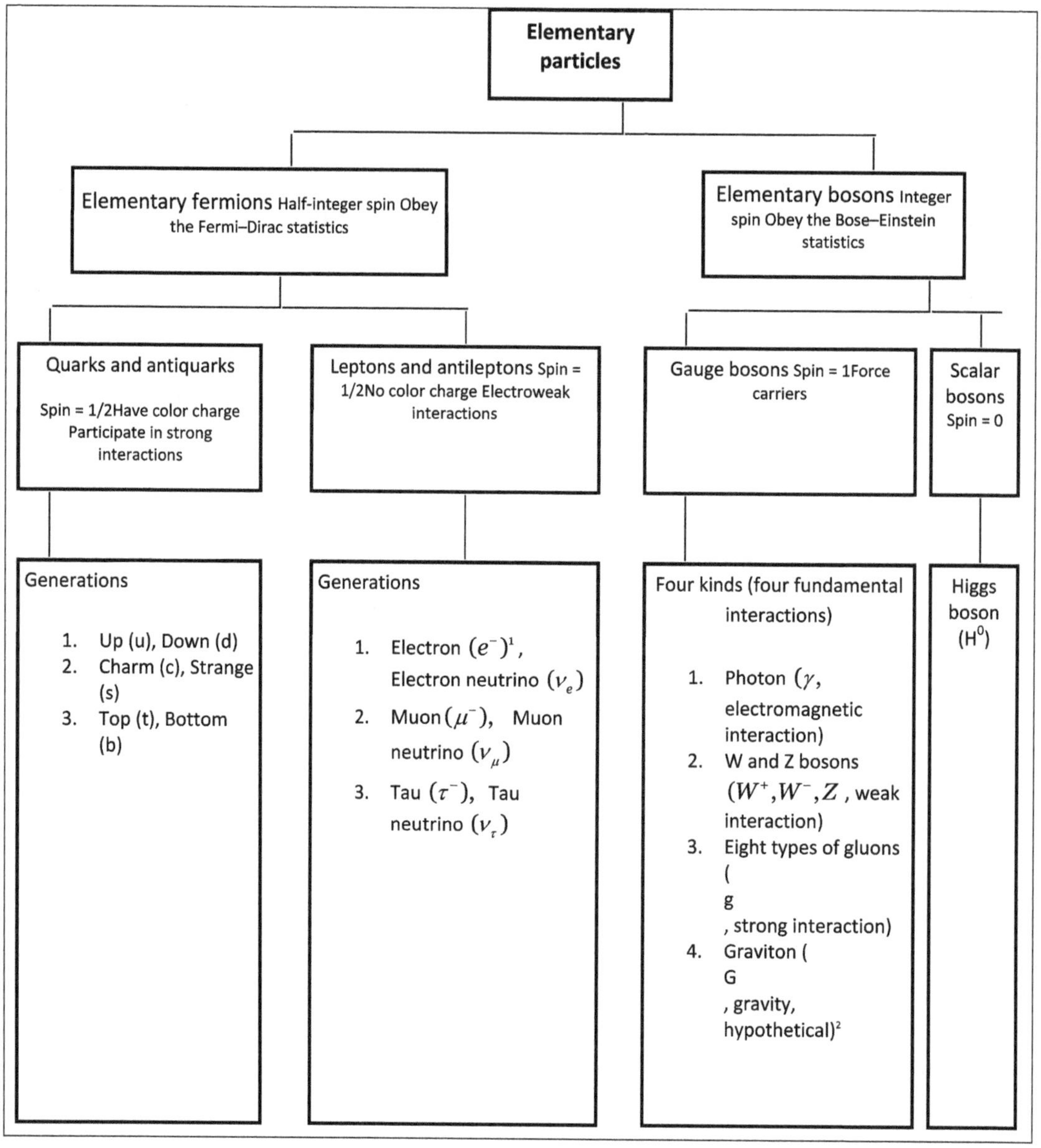

1. The antielectron (e^+) is traditionally called positron.

2. The known force carrier bosons all have spin = 1 and are therefore vector bosons. The hypothetical graviton has spin = 2 and is a tensor boson; whether it is a gauge boson as well, is unknown.

Fermions

The Standard Model includes 12 elementary particles of spin 1/2, known as fermions. According to the spin–statistics theorem, fermions respect the Pauli exclusion principle. Each fermion has a corresponding antiparticle.

The fermions of the Standard Model are classified according to how they interact (or equivalently, by what charges they carry). There are six quarks (up, down, charm, strange, top, bottom), and six leptons (electron, electron neutrino, muon, muon neutrino, tau, tau neutrino). Pairs from each classification are grouped together to form a generation, with corresponding particles exhibiting similar physical behavior.

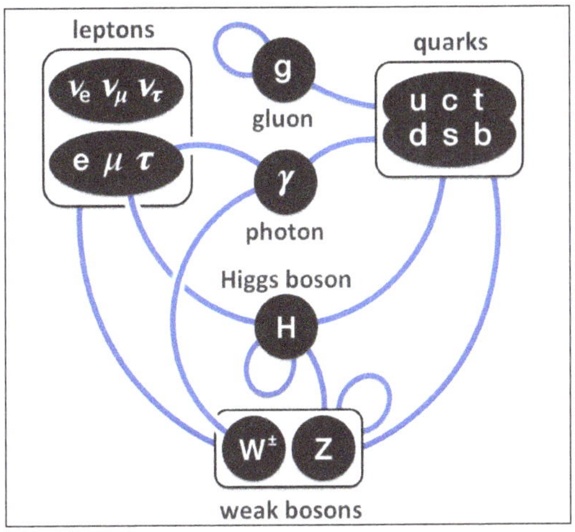

Summary of interactions between particles described by the Standard Model.

The defining property of the quarks is that they carry color charge, and hence interact via the strong interaction. A phenomenon called color confinement results in quarks being very strongly bound to one another, forming color-neutral composite particles (hadrons) containing either a quark and an antiquark (mesons) or three quarks (baryons). The familiar proton and neutron are the two baryons having the smallest mass. Quarks also carry electric charge and weak isospin. Hence they interact with other fermions both electromagnetically and via the weak interaction. The remaining six fermions do not carry color charge and are called leptons. The three neutrinos do not carry electric charge either, so their motion is directly influenced only by the weak nuclear force, which makes them notoriously difficult to detect. However, by virtue of carrying an electric charge, the electron, muon, and tau all interact electromagnetically.

Each member of a generation has greater mass than the corresponding particles of lower generations. The first-generation charged particles do not decay, hence all ordinary (baryonic) matter is made of such particles. Specifically, all atoms consist of electrons orbiting around atomic nuclei, ultimately constituted of up and down quarks. Second- and third-generation charged particles, on the other hand, decay with very short half-lives and are observed only in very high-energy environments. Neutrinos of all generations also do not decay and pervade the universe, but rarely interact with baryonic matter.

Gauge Bosons

In the Standard Model, gauge bosons are defined as force carriers that mediate the strong, weak, and electromagnetic fundamental interactions.

Interactions in physics are the ways that particles influence other particles. At a macroscopic level, electromagnetism allows particles to interact with one another via electric and magnetic fields, and gravitation allows particles with mass to attract one another in accordance with Einstein's theory of general relativity. The Standard Model explains such forces as resulting from matter particles exchanging other particles, generally referred to as *force mediating particles*. When a force-mediating particle is exchanged, at a macroscopic level the effect is equivalent to a force influencing both of them, and the particle is therefore said to have *mediated* (i.e., been the agent of) that force. The Feynman diagram calculations, which are a graphical representation of the perturbation theory approximation, invoke "force mediating particles", and when applied to analyze high-energy scattering experiments are in reasonable agreement with the data. However, perturbation theory (and with it the concept of a "force-mediating particle") fails in other situations. These include low-energy quantum chromodynamics, bound states, and solitons.

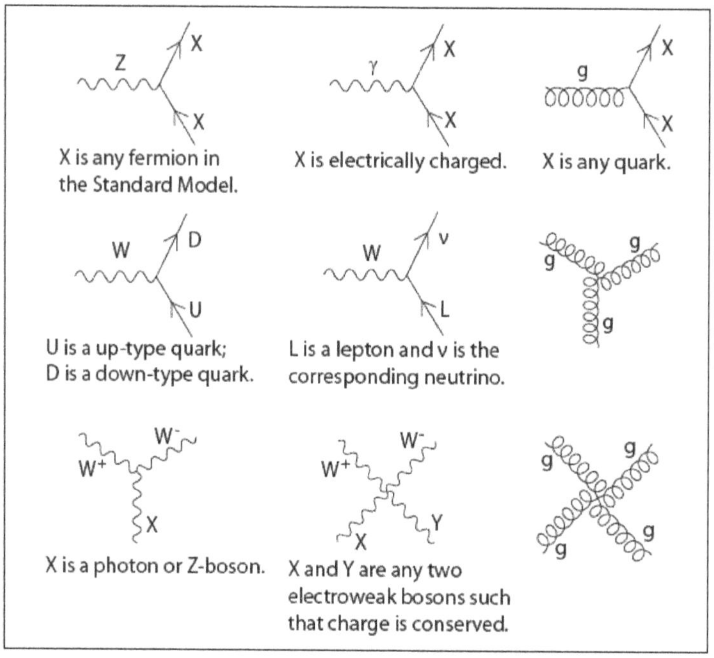

The above interactions form the basis of the standard model. Feynman diagrams in the standard model are built from these vertices. Modifications involving Higgs boson interactions and neutrino oscillations are omitted. The charge of the W bosons is dictated by the fermions they interact with; the conjugate of each listed vertex is also allowed.

The gauge bosons of the Standard Model all have spin (as do matter particles). The value of the spin is 1, making them bosons. As a result, they do not follow the Pauli exclusion principle that constrains fermions: thus bosons (e.g. photons) do not have a theoretical limit on their spatial density (number per volume). The different types of gauge bosons are described below.

- Photons mediate the electromagnetic force between electrically charged particles. The photon is massless and is well-described by the theory of quantum electrodynamics.
- The W^+, W^-, and Z gauge bosons mediate the weak interactions between particles of different flavors (all quarks and leptons). They are massive, with the Z being more massive than the $W^{\pm}$. The weak interactions involving the $W^{\pm}$ exclusively act on left-handed particles and right-handed antiparticles. Furthermore, the $W^{\pm}$ carries an electric charge of +1 and −1 and couples to the electromagnetic interaction. The electrically neutral Z boson interacts with both left-handed particles and antiparticles. These three gauge bosons along with the photons are grouped together, as collectively mediating the electroweak interaction.

- The eight gluons mediate the strong interactions between color charged particles (the quarks). Gluons are massless. The eightfold multiplicity of gluons is labeled by a combination of color and anticolor charge (e.g. red–antigreen). Because the gluons have an effective color charge, they can also interact among themselves. The gluons and their interactions are described by the theory of quantum chromodynamics.

The interactions between all the particles described by the Standard Model are summarized by the diagrams on the right of this section.

Higgs Boson

The Higgs particle is a massive scalar elementary particle theorized by Peter Higgs in 1964, when he showed that Goldstone's 1962 theorem (generic continuous symmetry, which is spontaneously broken) provides a third polarisation of a massive vector field. Hence, Goldstone's original scalar doublet, the massive spin-zero particle, was proposed as the Higgs boson. and is a key building block in the Standard Model. It has no intrinsic spin, and for that reason is classified as a boson (like the gauge bosons, which have integer spin).

The Higgs boson plays a unique role in the Standard Model, by explaining why the other elementary particles, except the photon and gluon, are massive. In particular, the Higgs boson explains why the photon has no mass, while the W and Z bosons are very heavy. Elementary-particle masses, and the differences between electromagnetism (mediated by the photon) and the weak force (mediated by the W and Z bosons), are critical to many aspects of the structure of microscopic (and hence macroscopic) matter. In electroweak theory, the Higgs boson generates the masses of the leptons (electron, muon, and tau) and quarks. As the Higgs boson is massive, it must interact with itself.

Because the Higgs boson is a very massive particle and also decays almost immediately when created, only a very high-energy particle accelerator can observe and record it. Experiments to confirm and determine the nature of the Higgs boson using the Large Hadron Collider (LHC) at CERN began in early 2010 and were performed at Fermilab's Tevatron until its closure in late 2011. Mathematical consistency of the Standard Model requires that any mechanism capable of generating the masses of elementary particles becomes visible at energies above 1.4 TeV; therefore, the LHC (designed to collide two 7 TeV proton beams) was built to answer the question of whether the Higgs boson actually exists.

On 4 July 2012, two of the experiments at the LHC (ATLAS and CMS) both reported independently that they found a new particle with a mass of about 125 GeV/c^2 (about 133 proton masses, on the order of 10×10^{-25} kg), which is "consistent with the Higgs boson". It was later confirmed to be the searched-for Higgs boson.

Theoretical Aspects

Construction of the Standard Model Lagrangian

Parameters of the Standard Model			
Symbol	Description	Renormalization scheme (point)	Value
m_e	Electron mass		511 keV

m_e	Muon mass		105.7 MeV
m_τ	Tau mass		1.78 GeV
m_u	Up quark mass	$\mu_{MS} = 2\ GeV$	1.9 MeV
m_d	Down quark mass	$\mu_{MS} = 2\ GeV$	4.4 MeV
m_s	Strange quark mass	$\mu_{MS} = 2\ GeV$	87 MeV
m_c	Charm quark mass	$\mu_{MS} = m_b$	1.32 GeV
m_c	Bottom quark mass	$\mu_{MS} = m_c$	4.24 GeV
m_t	Top quark mass	On shell scheme	173.5 GeV
θ_{12}	CKM 12-mixing angle		13.1°
θ_{13}	CKM 23-mixing angle		2.4°
θ_{23}	CKM 13-mixing angle		0.2°
δ	CKM CP violation Phase		0.995
$g_1$org′	U(1) gauge coupling	$\mu_{MS} = m_Z$	0.357
$g_2$org′	SU(2) gauge coupling	$\mu_{MS} = m_Z$	0.652
$g_3$org′	SU(3) gauge coupling	$\mu_{MS} = m_Z$	1.221
θ_{QCD}	QCD vacuum angle		~0
v	Higgs vacuum expectation value		246 GeV
m_H	Higgs mass		125.09±0.24 GeV

Technically,quantum fieldtheory provides the mathematical framework for the Standard Model, in which a Lagrangian controls the dynamics and kinematics of the theory. Each kind of particle is described in terms of a dynamical field that pervades space-time. The construction of the Standard Model proceeds following the modern method of constructing most field theories: by first postulating a set of symmetries of the system, and then by writing down the most general renormalizable Lagrangian from its particle (field) content that observes these symmetries.

The global Poincaré symmetry is postulated for all relativistic quantum field theories. It consists of the familiar translational symmetry, rotational symmetry and the inertial reference frame invariance central to the theory of special relativity. The local SU(3)×SU(2)×U(1) gauge symmetry is an internal symmetry that essentially defines the Standard Model. Roughly, the three factors of the gauge symmetry give rise to the three fundamental interactions. The fields fall into different representations of the various symmetry groups of the Standard Model Upon writing the most general

Lagrangian, one finds that the dynamics depends on 19 parameters, whose numerical values are established by experiment.

Quantum Chromodynamics Sector

The quantum chromodynamics (QCD) sector defines the interactions between quarks and gluons, which is a Yang–Mills gauge theory with SU(3) symmetry, generated by $1+(-1)+0=0$. Since leptons do not interact with gluons, they are not affected by this sector. The Dirac Lagrangian of the quarks coupled to the gluon fields is given by,

$$\mathcal{L}_{\mathrm{QCD}} = \sum_{\psi} \bar{\psi}_i \left(i\gamma^{\mu} (\partial_{\mu}\delta_{ij} - i g_s G^a_{\mu} T^a_{ij}) - m_{\psi}\delta_{ij} \right) \psi_j - \frac{1}{4} G^a_{\mu\nu} G^{\mu\nu}_a ,$$

where,

ψ_i is the Dirac spinor of the quark field, where $i = \{r, g, b\}$ represents color,

γ^{μ} are the Dirac matrices,

G^a_{μ} is the 8-component $(a=1,2,\ldots,8a=1,2,\ldots,8) SU(3)$ gauge field,

T^a_{ij} are the 3 × 3 Gell-Mann matrices, generators of the SU(3) color group,

$G^a_{\mu\nu}$ represents the gluon field strength tensor,

g_s is the strong coupling constant.

Electroweak Sector

The electroweak sector is a Yang–Mills gauge theory with the symmetry group U(1) × $SU(2)_L$,

$$\mathcal{L}_{\mathrm{EW}} = \sum_{\psi} \bar{\psi}\gamma^{\mu} \left(i\partial_{\mu} - g'\tfrac{1}{2} Y_{\mathrm{W}} B_{\mu} - g\tfrac{1}{2}\vec{\tau}_{\mathrm{L}} \vec{W}_{\mu} \right) \psi - \tfrac{1}{4} W_a^{\mu\nu} W^a_{\mu\nu} - \tfrac{1}{4} B^{\mu\nu} B_{\mu\nu}$$

where,

B_{μ} s the U(1) gauge field,

Y_W is the weak hypercharge – the generator of the U(1) group,

$W \rightarrow_{\mu}$ is the 3-component SU(2) gauge field,

τ_L are the Pauli matrices – infinitesimal generators of the SU(2) group – with subscript L to indicate that they only act on *left*-chiral fermions,

g' and *g* are the U(1) and SU(2) coupling constants respectively,

$W^{a\mu\nu} (a=1,2,3a=1,2,3)$ $B^{\mu\nu}$ and are the field strength tensors for the weak isospin and weak hypercharge fields.

Notice that the addition of fermion mass terms into the electroweak lagrangian is forbidden, since terms of the form do not respect U(1) × $SU(2)_L$ gauge invariance. Neither is it possible to add explicit mass terms for the U(1) and SU(2) gauge fields. The Higgs mechanism is responsible for the generation of the gauge boson masses, and the fermion masses result from Yukawa-type interactions with the Higgs field.

Higgs Sector

In the Standard Model, the Higgs field is a complex scalar of the group $SU(2)_L$:

$$\varphi = \frac{1}{\sqrt{2}}\begin{pmatrix}\varphi^+ \\ \varphi^0\end{pmatrix},$$

where the superscripts + and 0 indicate the electric charge (Q) of the components. The weak hypercharge (Y_W) of both components is 1.

Before symmetry breaking, the Higgs Lagrangian is

$$\mathcal{L}_H = \varphi^\dagger\left(\partial^\mu - \frac{i}{2}\left(g'Y_W B^\mu + g\vec{\tau}\vec{W}^\mu\right)\right)\left(\partial_\mu + \frac{i}{2}\left(g'Y_W B_\mu + g\vec{\tau}\vec{W}_\mu\right)\right)\varphi - \frac{\lambda^2}{4}\left(\varphi^\dagger\varphi - v^2\right)^2,$$

which up to a divergence term, (i.e. after partial integration) can also be written as

$$\mathcal{L}_H = \left|\left(\partial_\mu + \frac{i}{2}\left(g'Y_W B_\mu + g\vec{\tau}\vec{W}_\mu\right)\right)\varphi\right|^2 - \frac{\lambda^2}{4}\left(\varphi^\dagger\varphi - v^2\right)^2.$$

Yukawa Sector

The Yukawa interaction terms are:

$$\mathcal{L}_{\text{Yukawa}} = \bar{U}_L G_u U_R \phi^0 - \bar{D}_L G_u U_R \phi^- + \bar{U}_L G_d D_R \phi^+ + \bar{D}_L G_d D_R \phi^0 + hc.$$

where $G_{u,d}$ are 3 × 3 matrices of Yukawa couplings, with the ij term giving the coupling of the generations i and j.

The Standard Model describes three of the four fundamental forces in nature; only gravity remains unexplained. In the Standard Model, a force is described as an exchange of bosons between the objects affected, such as a photon for the electromagnetic force and a gluon for the strong interaction. Those particles are called force carriers or messenger particles.

The four fundamental interactions of nature					
Property/Interaction	Gravitation	Weak	Electromagnetic	Strong	
		(Electroweak)		Fundamental	Residual
Mediating particles	Not yet observed (Graviton hypothesised)	W^+, W^- and Z^0	γ (photon)	Gluons	π, ρ and ω mesons

Affected particles	All particles	Left-handed fermions	Electrically charged	Quarks, Gluons	Hadrons
Acts on	Mass, Energy	Flavor	Electric charge	Color charge	
Bound states formed	Planets, Stars, Solar systems, Galaxies	n/a	Atoms, Molecules	Hadrons	Atomic nuclei
Strength at the scale of quarks (relative to electromagnetism)	10^{-41} (predicted)	10^{-4}	1	60	Not applicable to quarks
Strength at the scale of protons/neutrons (relative to electromagnetism)	10^{-36} (predicted)	10^{-7}	1	Not applicable to hadrons	20

Tests and Predictions

The Standard Model (SM) predicted the existence of the W and Z bosons, gluon, and the top and charm quarks and predicted many of their properties before these particles were observed. The predictions were experimentally confirmed with good precision.

The SM also predicted the existence of the Higgs boson, found in 2012 at the Large Hadron Collider, as the last particle of the SM.

Challenges

Self-consistency of the Standard Model (currently formulated as a non-abelian gauge theory quantized through path-integrals) has not been mathematically proven. While regularized versions useful for approximate computations (for example lattice gauge theory) exist, it is not known whether they converge (in the sense of S-matrix elements) in the limit that the regulator is removed. A key question related to the consistency is the Yang–Mills existence and mass gap problem.

Experiments indicate that neutrinos have mass, which the classic Standard Model did not allow. To accommodate this finding, the classic Standard Model can be modified to include neutrino mass.

If one insists on using only Standard Model particles, this can be achieved by adding a non-renormalizable interaction of leptons with the Higgs boson. On a fundamental level, such an interaction emerges in the seesaw mechanism where heavy right-handed neutrinos are added to the theory. This is natural in the left-right symmetric extension of the Standard Model and in certain grand unified theories. As long as new physics appears below or around 10^{14} GeV, the neutrino masses can be of the right order of magnitude.

Theoretical and experimental research has attempted to extend the Standard Model into a Unified field theory or a Theory of everything, a complete theory explaining all physical phenomena including constants. Inadequacies of the Standard Model that motivate such research include:

- The model does not explain gravitation, although physical confirmation of a theoretical particle known as a graviton would account for it to a degree. Though it addresses strong and electroweak interactions, the Standard Model does not consistently explain the canonical theory of gravitation, general relativity, in terms of quantum field theory. The reason for this is, among other things, that quantum field theories of gravity generally break down before reaching the Planck scale. As a consequence, we have no reliable theory for the very early universe.

- Some physicists consider it to be *ad hoc* and inelegant, requiring 19 numerical constants whose values are unrelated and arbitrary. Although the Standard Model, as it now stands, can explain why neutrinos have masses, the specifics of neutrino mass are still unclear. It is believed that explaining neutrino mass will require an additional 7 or 8 constants, which are also arbitrary parameters.
- The Higgs mechanism gives rise to the hierarchy problem if some new physics (coupled to the Higgs) is present at high energy scales. In these cases, in order for the weak scale to be much smaller than the Planck scale, severe fine tuning of the parameters is required; there are, however, other scenarios that include quantum gravity in which such fine tuning can be avoided. There are also issues of quantum triviality, which suggests that it may not be possible to create a consistent quantum field theory involving elementary scalar particles.
- The model is inconsistent with the emerging Lambda-CDM model of cosmology. Contentions include the absence of an explanation in the Standard Model of particle physics for the observed amount of cold dark matter (CDM) and its contributions to dark energy, which are many orders of magnitude too large. It is also difficult to accommodate the observed predominance of matter over antimatter (matter/antimatter asymmetry). The isotropy and homogeneity of the visible universe over large distances seems to require a mechanism like cosmic inflation, which would also constitute an extension of the Standard Model.

Weak Interaction

In particle physics, the weak interaction, which is also often called the weak force or weak nuclear force, is the mechanism of interaction between subatomic particles that is responsible for the radioactive decay of atoms. The weak interaction serves an essential role in nuclear fission, and the theory regarding it in terms of both its behavior and effects is sometimes called quantum flavordynamics (QFD). However, the term QFD is rarely used because the weak force is better understood in terms of electroweak theory (EWT). In addition to this, QFD is related to quantum chromodynamics (QCD), which deals with the strong interaction, and quantum electrodynamics (QED), which deals with the electromagnetic force.

The effective range of the weak force is limited to subatomic distances, and is less than the diameter of a proton. It is one of the four known force-related fundamental interactions of nature, alongside the strong interaction, electromagnetism, and gravitation.

Properties

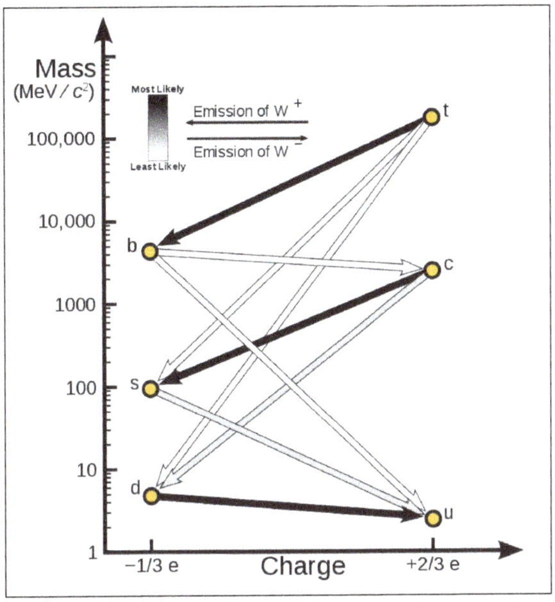

A diagram depicting the various decay routes due to the weak interaction and some indication of their likelihood. The intensity of the lines is given by the CKM parameters.

The weak interaction is unique in a number of respects:

- It is the only interaction capable of changing the flavor of quarks (i.e., of changing one type of quark into another).
- It is the only interaction that violates P or parity-symmetry. It is also the only one that violates charge-parity CP symmetry.
- It is mediated (propagated) by force carrier particles that have significant masses, an unusual feature which is explained in the Standard Model by the Higgs mechanism.

Due to their large mass (approximately 90 GeV/c^2) these carrier particles, termed the W and Z bosons, are short-lived with a lifetime of under 10^{-24} seconds. The weak interaction has a coupling constant (an indicator of interaction strength) of between 10^{-7} and 10^{-6}, compared to the strong interaction's coupling constant of 1 and the electromagnetic coupling constant of about 10^{-2}; consequently the weak interaction is weak in terms of strength. The weak interaction has a very short effective range (around 10^{-17} to 10^{-16} m). At distances around 10^{-18} meters, the weak interaction has a strength of a similar magnitude to the electromagnetic force, but this starts to decrease exponentially with increasing distance. Scaled up by just one and a half orders of magnitude, at distances of around 3×10^{-17} m, the weak interaction becomes 10,000 times weaker.

The weak interaction affects all the fermions of the Standard Model, as well as the Higgs boson; neutrinos interact through gravity and the weak interaction only, and neutrinos were the original reason for the name *weak force*. The weak interaction does not produce bound states nor does it involve binding energy – something that gravity does on an astronomical scale, that the electromagnetic force does at the atomic level, and that the strong nuclear force does inside nuclei.

Its most noticeable effect is due to its first unique feature: flavor changing. A neutron, for example, is heavier than a proton (its sister nucleon), but it cannot decay into a proton without changing the flavor (type) of one of its two *down* quarks to an *up* quark. Neither the strong interaction nor electromagnetism permit flavor changing, so this proceeds by weak decay; without weak decay, quark properties such as strangeness and charm (associated with the quarks of the same name) would also be conserved across all interactions.

All mesons are unstable because of weak decay. In the process known as beta decay, a *down* quark in the neutron can change into an *up* quark by emitting a virtual W^-boson which is then converted into an electron and an electron antineutrino. Another example is the electron capture, a common variant of radioactive decay, wherein a proton and an electron within an atom interact, and are changed to a neutron (an up quark is changed to a down quark) and an electron neutrino is emitted.

Due to the large masses of the W bosons, particle transformations or decays (e.g., flavor change) that depend on the weak interaction typically occur much more slowly than transformations or decays that depend only on the strong or electromagnetic forces. For example, a neutral pion decays electromagnetically, and so has a life of only about 10^{-16} seconds. In contrast, a charged pion can only decay through the weak interaction, and so lives about 10^{-8} seconds, or a hundred million times longer than a neutral pion. A particularly extreme example is the weak-force decay of a free neutron, which takes about 15 minutes.

Weak Isospin and Weak Hypercharge

Left-handed fermions in the Standard Model								
Generation 1			Generation 2			Generation 3		
Fermion	Symbol	Weak isospin	Fermion	Symbol	Weak isospin	Fermion	Symbol	Weak isospin
Electron neutrino	ν_e	$+\frac{1}{2}$	Muon neutrino	ν_μ	$+\frac{1}{2}$	Tau neutrino	ν_τ	$+\frac{1}{2}$
Electron	e^-	$-\frac{1}{2}$	Muon	μ^-	$-\frac{1}{2}$	Tau	τ^-	$-\frac{1}{2}$
Up quark	u	$+\frac{1}{2}$	Charm quark	c	$+\frac{1}{2}$	Top quark		–
Down quark	d	$-\frac{1}{2}$	Strange quark	s	$-\frac{1}{2}$	Bottom quark	b	$-\frac{1}{2}$
All of the above left-handed (*regular*) particles have corresponding right-handed *anti*-particles with equal and opposite weak isospin.								
All right-handed (regular) particles and left-handed antiparticles have weak isospin of 0.								

All particles have a property called *weak isospin* (symbol T_3), which serves as a quantum number and governs how that particle behaves in the weak interaction. Weak isospin plays the same role in the weak interaction as does electric charge in electromagnetism, and color charge in the strong interaction. All left-handed fermions have a weak isospin value of either $+{}^{1}/_{2}$*or* $-{}^{1}/_{2}$. For example, the up quark has a $T_3 -{}^{1}/_{2}$ and the down quark $-\frac{1}{2}$. A quark never decays through the weak interaction into a quark of the same T_3: Quarks with a $T_3 -{}^{1}/_{2}$ only decay into quarks with a T_3 of $-{}^{1}/_{2}$ and vice versa.

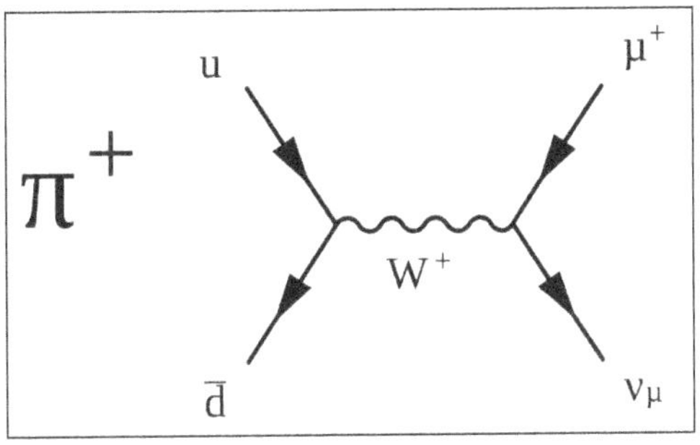

π^{+} decay through the weak interaction.

In any given interaction, weak isospin is conserved: The sum of the weak isospin numbers of the particles entering the interaction equals the sum of the weak isospin numbers of the particles exiting that interaction. For example, a (left-handed) π^{+}, with a weak isospin of +1 normally decays into a ν_{μ} ($+^{1}/_{2}$) and a μ^{+} (as a right-handed antiparticle, $+^{1}/_{2}$).

Following the development of the electroweak theory, another property, weak hypercharge, was developed. It is dependent on a particle's electrical charge and weak isospin, and is defined by:

$$Y_{\mathrm{W}} = 2(Q - T_{3})$$

where Y_{W} is the weak hypercharge of a given type of particle, Q is its electrical charge (in elementary charge units) and T_{3} is its weak isospin. Whereas some particles have a weak isospin of zero, all spin-$^{1}/_{2}$ particles have non-zero weak hypercharge. Weak hypercharge is the generator of the U(1) component of the electroweak gauge group.

Interaction Types

There are two types of weak interaction (called *vertices*). The first type is called the "charged-current interaction" because it is mediated by particles that carry an electric charge (the W^{+}or W^{-} bosons), and is responsible for the beta decay phenomenon. The second type is called the "neutral-current interaction" because it is mediated by a neutral particle, the Z boson.

Charged-current Interaction

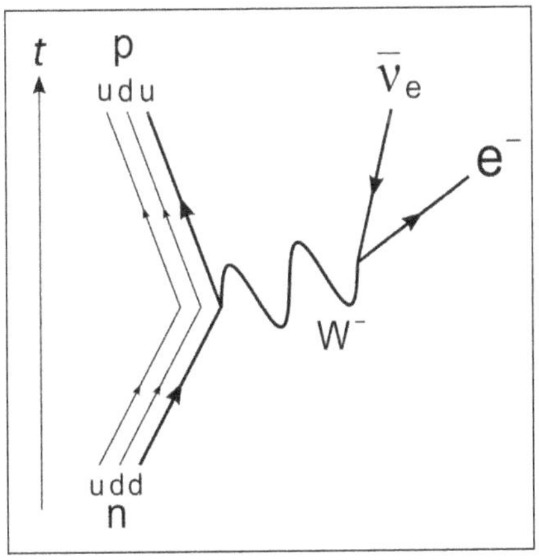

The Feynman diagram for beta-minus decay of a neutron into a proton, electron and electron anti-neutrino, via an intermediate heavy W^{-} boson.

In one type of charged current interaction, a charged lepton (such as an electron or a muon, having a charge of −1) can absorb a W^+boson (a particle with a charge of +1) and be thereby converted into a corresponding neutrino (with a charge of 0), where the type ("flavor") of neutrino (electron, muon or tau) is the same as the type of lepton in the interaction, for example:

$$\mu^- + W^+ \rightarrow \nu_\mu$$

Similarly, a down-type quark (d with a charge of $-^1/_3$) can be converted into an up-type quark (u, with a charge of $+^2/_3$), by emitting a W^- boson or by absorbing a W^+ boson. More precisely, the down-type quark becomes a quantum superposition of up-type quarks: that is to say, it has a possibility of becoming any one of the three up-type quarks, with the probabilities given in the CKM matrix tables. Conversely, an up-type quark can emit a W^+boson, or absorb a W^-boson, and thereby be converted into a down-type quark, for example:

$$d \rightarrow u + W^-$$
$$d + W^+ \rightarrow u$$
$$c \rightarrow s + W^+$$
$$c + W^- \rightarrow s$$

The W boson is unstable so will rapidly decay, with a very short lifetime. For example:

$$W^- \rightarrow e^- + \bar{\nu}_e$$
$$W^+ \rightarrow e^+ + \nu_e$$

Decay of the W^+ boson to other products can happen, with varying probabilities.

In the so-called beta decay of a neutron a down quark within the neutron emits a virtual W^-boson and is thereby converted into an up quark, converting the neutron into a proton. Because of the energy involved in the process (i.e., the mass difference between the down quark and the up quark), the W^- boson can only be converted into an electron and an electron-antineutrino. At the quark level, the process can be represented as:

$$d \rightarrow u + e^- + \bar{\nu}_e$$

Neutral-current Interaction

In neutral current interactions, a quark or a lepton (e.g., an electron or a muon) emits or absorbs a neutral Z^0 boson. For example:

$$e^- \rightarrow e^- + Z^0$$

Like the W^+ boson, the Z^0 boson also decays rapidly, for example:

$$Z^0 \rightarrow b + \bar{b}$$

Electroweak Theory

The Standard Model of particle physics describes the electromagnetic interaction and the weak interaction as two different aspects of a single electroweak interaction. This theory was developed around 1968 by Sheldon Glashow, Abdus Salam and Steven Weinberg, and they were awarded the 1979 Nobel Prize in Physics for their work. The Higgs mechanism provides an explanation for the presence of three massive gauge bosons (W^+, W^-, Z^o, the three carriers of the weak interaction) and the massless photon (γ, the carrier of the electromagnetic interaction).

According to the electroweak theory, at very high energies, the universe has four components of the Higgs field whose interactions are carried by four massless gauge bosons – each similar to the photon – forming a complex scalar Higgs field doublet. However, at low energies, this gauge symmetry is spontaneously broken down to the **U**(1) symmetry of electromagnetism, since one of the Higgs fields acquires a vacuum expectation value. This symmetry-breaking would be expected to produce three massless bosons, but instead they become integrated by the other three fields and acquire mass through the Higgs mechanism. These three boson integrations produce the W^+, W^- *and* Z^o *b* bosons of the weak interaction. The fourth gauge boson is the photon of electromagnetism, and remains massless.

This theory has made a number of predictions, including a prediction of the masses of the Z and W-bosons before their discovery. On 4 July 2012, the CMS and the ATLAS experimental teams at the Large Hadron Collider independently announced that they had confirmed the formal discovery of a previously unknown boson of mass between 125–127 GeV/c^2, whose behaviour so far was "consistent with" a Higgs boson, while adding a cautious note that further data and analysis were needed before positively identifying the new boson as being a Higgs boson of some type. By 14 March 2013, the Higgs boson was tentatively confirmed to exist.

If the electroweak symmetry breaking scale were lowered, the unbroken SU(2) interaction would eventually become confining. Alternative models where SU(2) becomes confining above that scale are quantitatively similar to the Standard Model at lower energies, but dramatically different above symmetry breaking.

Violation of Symmetry

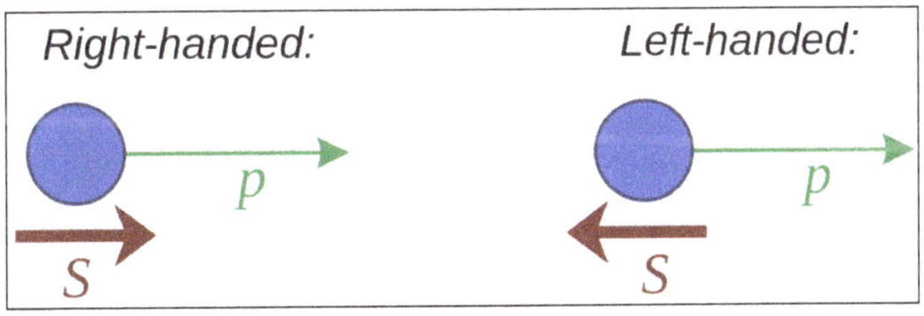

Left- and right-handed particles: p is the particle's momentum and S is its spin.
Note the lack of reflective symmetry between the states.

The laws of nature were long thought to remain the same under mirror reflection. The results of an experiment viewed via a mirror were expected to be identical to the results of a mirror-reflected copy of the experimental apparatus. This so-called law of parity conservation was known to be respected by classical gravitation, electromagnetism and the strong interaction; it was assumed to be a universal law. However, in the mid-1950s Chen-Ning Yang and Tsung-Dao Lee suggested that the weak interaction might violate this law. Chien Shiung Wu and collaborators in 1957 discovered that the weak interaction violates parity, earning Yang and Lee the 1957 Nobel Prize in Physics.

Although the weak interaction was once described by Fermi's theory, the discovery of parity violation and renormalization theory suggested that a new approach was needed. In 1957, Robert Marshak and George Sudarshan and, somewhat later, Richard Feynman and Murray Gell-Mann proposed a V–A (vector minus axial vector or left-handed) Lagrangian for weak interactions. In this theory, the weak interaction acts only on left-handed particles (and right-handed antiparticles). Since the mirror reflection of a left-handed particle is right-handed, this explains the maximal violation of parity. The V–A theory was developed before the discovery of the Z boson, so it did not include the right-handed fields that enter in the neutral current interaction.

However, this theory allowed a compound symmetry CP to be conserved. CP combines parity P (switching left to right) with charge conjugation C (switching particles with antiparticles). Physicists were again surprised when in 1964, James Cronin and Val Fitch provided clear evidence in kaon decays that CP symmetry could be broken too, winning them the 1980 Nobel Prize in Physics. In 1973, Makoto Kobayashi and Toshihide Maskawa showed that CP violation in the weak interaction required more than two generations of particles, effectively predicting the existence of a then unknown third generation. This discovery earned them half of the 2008 Nobel Prize in Physics.

Unlike parity violation, CP violation occurs only in limited circumstances. Despite its rarity, it is widely believed to be the reason that there is much more matter than antimatter in the universe, and thus forms one of Andrei Sakharov's three conditions for baryogenesis.

Particle Conservation Laws

Conservation laws are critical to an understanding of particle physics. Strong evidence exists that energy, momentum, and angular momentum are all conserved in all particle interactions. The annihilation of an electron and positron at rest, for example, cannot produce just one photon because this violates the conservation of linear momentum, the special theory of relativity modifies definitions of momentum, energy, and other familiar quantities. In particular, the relativistic momentum of a particle differs from its classical momentum by a factor $\gamma = 1/\sqrt{1-(v/c)^2}$ that varies from 1 to ∞, depending on the speed of the particle.

charge is conserved in all electrostatic phenomena. Charge lost in one place is gained in another because charge is carried by particles. No known physical processes violate charge conservation. In the next section, we describe three less-familiar conservation laws: baryon number, lepton number, and strangeness. These are by no means the only conservation laws in particle physics.

Baryon Number Conservation

No conservation law considered thus far prevents a neutron from decaying via a reaction such as:

$$n \rightarrow e^+ + e^- .$$

This process conserves charge, energy, and momentum. However, it does not occur because it violates the law of baryon number conservation. This law requires that the total baryon number of a reaction is the same before and after the reaction occurs. To determine the total baryon number, every elementary particle is assigned a baryon number B. The baryon number has the value $B=+1$ for baryons, −1 for antibaryons, and 0 for all other particles. Returning to the above case (the decay of the neutron into an electron-positron pair), the neutron has a value $=+$, whereas the electron and the positron each has a value of 0. Thus, the decay does not occur because the total baryon number changes from 1 to 0. However, the proton-antiproton collision process,

$$p+\overline{p}\rightarrow p+p\overline{p}+\overline{p}$$

does satisfy the law of conservation of baryon number because the baryon number is zero before and after the interaction. The baryon number for several common particles is given in table.

Table: Conserved properties of particles.

Particle name	Symbol	Lepton number (L_e)	Lepton number (L_μ)	Lepton number (L_τ)	Baryon number (B)	Strangeness number
Electron	$e-$	1	0	0	0	0
Electron neutrino	ν_e	1	0	0	0	0
Muon	μ^-	0	1	0	0	0
Muon neutrino	ν_μ	0	1	0	0	0
Tau	τ−	0	0	1	0	0
Tau neutrino	ν_τ	0	0	1	0	0
Pion	π^+	0	0	0	0	0
Positive kaon	K^+	0	0	0	0	1
Negative kaon	K−	0	0	0	0	−1

Proton	p	0	0	0	1	0
Neutron	n	0	0	0	1	0
Lambda zero	Λo	0	0	0	1	−1
Positive sigma	Σ^+	0	0	0	1	−1
Negative sigma	Σ^-	0	0	0	1	−1
Xi zero	Ξ^0	0	0	0	1	−2
Negative xi	Ξ^-	0	0	0	1	−2
Omega	Ω^-	0	0	0	1	−3

Example:

Based on the law of conservation of baryon number, which of the following reactions can occur?

$$(a)\pi^- + p \rightarrow \pi^0 + n + \pi^- + \pi^+$$

$$(b)p + \overline{p} \rightarrow p + p + \overline{p}$$

Strategy

Determine the total baryon number for the reactants and products, and require that this value does not change in the reaction.

Solution:

For reaction (a), the net baryon number of the two reactants is $0+1=1$ and the net baryon number of the four products is $0+1+0+0=1$.

Since the net baryon numbers of the reactants and products are equal, this reaction is allowed on the basis of the baryon number conservation law.

For reaction (b), the net baryon number of the reactants is $1+(-1)=0$ and the net baryon number of the proposed products is $1+1+1+(-1)=1$. Since the net baryon numbers of the reactants and proposed products are not equal, this reaction cannot occur.

Significance

Baryon number is conserved in the first reaction, but not in the second. Baryon number conservation constrains what reactions can and cannot occur in nature.

Lepton Number Conservation

Lepton number conservation states that the sum of lepton numbers before and after the interaction must be the same. There are three different lepton numbers: the electron-lepton number L_e, the muon-lepton number L_μ, and the tau-lepton number L_τ. In any interaction, each of these quantities must be conserved separately. For electrons and electron neutrinos, $L_e = 1$; for their antiparticles $L_e = 1$; all other particles have $L_e = 0$. Similarly, $L_\mu = 1$ for muons and muon neutrinos $L_\mu = 1$ for their antiparticles, and $L_\mu = 0$ for all other particles. Finally $L_\tau = 1, -1$ or 0, depending on whether we have a tau or tau neutrino, their antiparticles, or any other particle, respectively. Lepton number conservation guarantees that the number of electrons and positrons in the universe stays relatively constant.

To illustrate the lepton number conservation law, consider the following known two-step decay process:

$$\pi^+ \to \mu^+ + \nu_\mu$$

$$\mu^+ \to e^+ + \nu_e + \overline{\nu}_\mu$$

In the first decay, all of the lepton numbers for π^+ are 0. For the products of this decay, $L_\mu = -1$ for μ^+ and $L_\mu = -1$ for ν_μ. Therefore, muon-lepton number is conserved. Neither electrons nor tau are involved in this decay, so $L_e = 0$ and $L_\tau = 0$ for the initial particle and all decay products. Thus, electron-lepton and tau-lepton numbers are also conserved. In the second decay, μ^+ has a muon-lepton number $L_\mu = -1$, whereas the net muon-lepton number of the decay products is $0 + 0 + (-1) = -1$. Thus, the muon-lepton number is conserved. Electron-lepton number is also conserved, as $L_\tau = 0$ for μ^+, whereas the net electron-lepton number of the decay products is $(-1)+1+0=0$. Finally, since no taus or tau-neutrons are involved in this decay, the tau-lepton number is also conserved.

Example:

Based on the law of conservation of lepton number, which of the following decays can occur?

$$(a) n \to p + e^- + \overline{\nu}_e$$

$$(b) \pi^- \to \mu^- + \nu_\mu + \nu_\mu$$

Strategy

Determine the total lepton number for the reactants and products, and require that this value does not change in the reaction.

Solution:

For decay (a), the electron-lepton number of the neutron is 0, and the net electron-lepton number of the decay products is:

$$0+1+(-1)=0$$

Since the net electron-lepton numbers before and after the decay are the same, the decay is possible on the basis of the law of conservation of electron-lepton number. Also, since there are no muons or taus involved in this decay, the muon-lepton and tauon-lepton numbers are conserved.

For decay (b), the muon-lepton number of the $\pi-$ is 0, and the net muon-lepton number of the proposed decay products is $1+1+(-1)=1$.

Thus, on the basis of the law of conservation of muon-lepton number, this decay cannot occur.

Significance

Lepton number is conserved in the first reaction, but not in the second. Lepton number conservation constrains what reactions can and cannot occur in nature.

Strangeness Conservation

In the late 1940s and early 1950s, cosmic-ray experiments revealed the existence of particles that had never been observed on Earth. These particles were produced in collisions of pions with protons or neutrons in the atmosphere. Their production and decay were unusual. They were produced in the strong nuclear interactions of pions and nucleons, and were therefore inferred to be hadrons; however, their decay was mediated by the much more slowly acting weak nuclear interaction. Their lifetimes were on the order of $10-^{10}$ to $10-^{8}\ s$, whereas a typical lifetime for a particle that decays via the strong nuclear reaction is $10\ ^{23}$. These particles were also unusual because they were always produced in pairs in the pion-nucleon collisions. For these reasons, these newly discovered particles were described as strange. The production and subsequent decay of a pair of strange particles is illustrated in figure and follows the reaction:

$$\pi^- + p \rightarrow \Lambda^0 + K^0$$

The lambda particle then decays through the weak nuclear interaction according to,

$$\Lambda^0 \rightarrow \pi^- + p,$$

and the kaon decays via the weak interaction,

$$K^0 \rightarrow \pi^+ + \pi^-.$$

To rationalize the behavior of these strange particles, particle physicists invented a particle property conserved in strong interactions but not in weak interactions. This property is called strangeness and, as the name suggests, is associated with the presence of a strange quark. The strangeness of a particle is equal to the number of strange quarks of the particle. Strangeness conservation requires the total strangeness of a reaction or decay (summing the strangeness of all the particles) is the same before and after the interaction. Strangeness conservation is not absolute: It is conserved in strong interactions and electromagnetic interactions but not in weak inter actions. The strangeness number for several common particles is given in table.

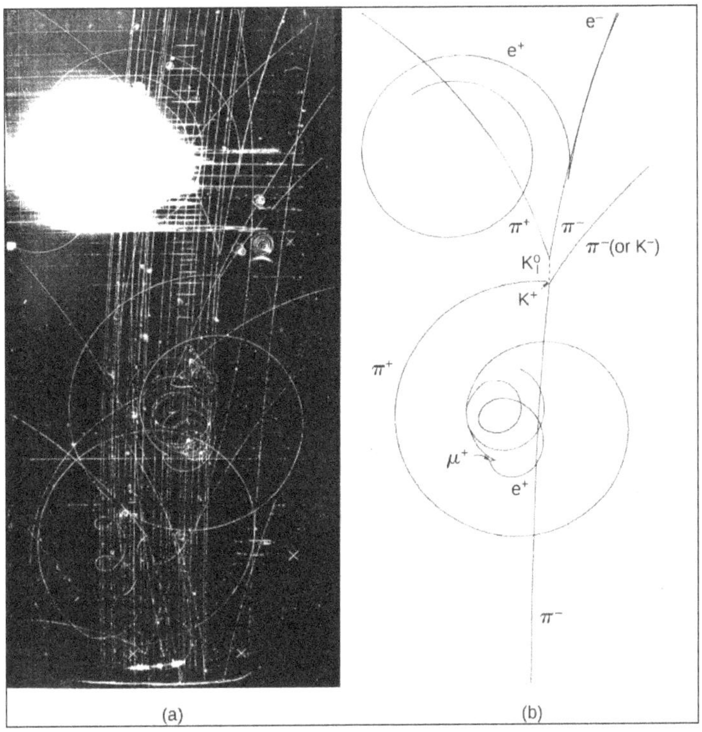

The interactions of hadrons. (a) Bubble chamber photograph; (b) sketch that represents the photograph.

Example:

Based on the conservation of strangeness, can the following reaction occur?

$$\pi^- + p \rightarrow K^+ + K^- + n.$$

The following decay is mediated by the weak nuclear force

$$K^+ \rightarrow \pi^+ + \pi^0.$$

Does the decay conserve strangeness? If not, can the decay occur?

Strategy

Determine the strangeness of the reactants and products and require that this value does not change in the reaction.

Solution:

The net strangeness of the reactants is $0+0=0$, and the net strangeness of the products is $1+(-1)+0=0$.

Thus, the strong nuclear interaction between a pion and a proton is not forbidden by the law of conservation of strangeness. Notice that baryon number is also conserved in the reaction.

The net strangeness before and after this decay is 1 and 0, so the decay does not conserve strangeness. However, the decay may still be possible, because the law of conservation of strangeness does not apply to weak decays.

Significance

Strangeness is conserved in the first reaction, but not in the second. Strangeness conservation constrains what reactions can and cannot occur in nature.

References

- What-is-a-particle-definition-theory-quiz: study.com, Retrieved 25 June, 2019
- Particle-classification: electron6.phys.utk.edu, Retrieved 08 May, 2019
- Four-fundamental-forces-of-nature: clearias.com, Retrieved 23 January, 2019
- Particle-physics: sciencedaily.com, Retrieved 01 August, 2019
- Howard Georgi (2007). "Unparticle Physics". Physical Review Letters. 98 (22): 221601. arXiv:hep-ph/0703260. Bibcode:2007PhRvL..98v1601G. doi:10.1103/PhysRevLett.98.221601. PMID 17677831
- R. Mann (2010). An Introduction to Particle Physics and the Standard Model. CRC Press. ISBN 978-1-4200-8298-2

Elementary Particles 2

- **Fermions**
- **Bosons**
- **The Quark Model**
- **Flavour Particle Physics**
- **Superpartner**

The subatomic particles with no sub structures are known as elementary particles. Fermions and bosons are the two primary types of elementary particles. This chapter has been carefully written to provide an easy understanding of these types and related concepts of elementary particles.

Elementary particle also called subatomic particle, is any of various self-contained units of matter or energy that are the fundamental constituents of all matter. Subatomic particles include electrons, the negatively charged, almost massless particles that nevertheless account for most of the size of the atom, and they include the heavier building blocks of the small but very dense nucleus of the atom, the positively charged protons and the electrically neutral neutrons. But these basic atomic components are by no means the only known subatomic particles. Protons and neutrons, for instance, are themselves made up of elementary particles called quarks, and the electron is only one member of a class of elementary particles that also includes the muon and the neutrino. More-unusual subatomic particles—such as the positron, the antimatter counterpart of the electron—have been detected and characterized in cosmic ray interactions in Earth's atmosphere. The field of subatomic particles has expanded dramatically with the construction of powerful particle accelerators to study high-energy collisions of electrons, protons, and other particles with matter. As particles collide at high energy, the collision energy becomes available for the creation of subatomic particles such as mesons and hyperons. Finally, completing the revolution that began in the early 20th century with theories of the equivalence of matter and energy, the study of subatomic particles has been transformed by the discovery that the actions of forces are due to the exchange of "force" particles such as photons and gluons. More than 200 subatomic particles have been detected—most of them highly unstable, existing for less than a millionth of a second—as a result of collisions produced in cosmic ray reactions or particle accelerator experiments. Theoretical and experimental research in particle physics, the study of subatomic particles and their properties, has given scientists a clearer understanding of the nature of matter and energy and of the origin of the universe.

The current understanding of the state of particle physics is integrated within a conceptual framework known as the Standard Model. The Standard Model provides a classification scheme for all the known subatomic particles based on theoretical descriptions of the basic forces of matter.

Basic Concepts of Particle Physics

Divisible Atom

The physical study of subatomic particles became possible only during the 20th century, with the development of increasingly sophisticated apparatuses to probe matter at scales of 10–15 metre and less (that is, at distances comparable to the diameter of the proton or neutron). Yet the basic philosophy of the subject now known as particle physics dates to at least 500 BCE, when the Greek philosopher Leucippus and his pupil Democritus put forward the notion that matter consists of invisibly small, indivisible particles, which they called atoms. For more than 2,000 years the idea of atoms lay largely neglected, while the opposing view that matter consists of four elements—earth, fire, air, and water—held sway. But by the beginning of the 19th century the atomic theory of matter had returned to favour, strengthened in particular by the work of John Dalton, an English chemist whose studies suggested that each chemical element consists of its own unique kind of atom. As such, Dalton's atoms are still the atoms of modern physics. By the close of the century, however, the first indications began to emerge that atoms are not indivisible, as Leucippus and Democritus had imagined, but that they instead contain smaller particles.

In 1896 the French physicist Henri Becquerel discovered radioactivity, and in the following year J.J. Thomson, a professor of physics at the University of Cambridge in England, demonstrated the existence of tiny particles much smaller in mass than hydrogen, the lightest atom. Thomson had discovered the first subatomic particle, the electron. Six years later Ernest Rutherford and Frederick Soddy, working at McGill University in Montreal, found that radioactivity occurs when atoms of one type transmute into those of another kind. The idea of atoms as immutable, indivisible objects had become untenable.

The basic structure of the atom became apparent in 1911, when Rutherford showed that most of the mass of an atom lies concentrated at its centre, in a tiny nucleus. Rutherford postulated that the atom resembled a miniature solar system, with light, negatively charged electrons orbiting the dense, positively charged nucleus, just as the planets orbit the Sun. The Danish theorist Niels Bohr refined this model in 1913 by incorporating the new ideas of quantization that had been developed by the German physicist Max Planck at the turn of the century. Planck had theorized that electromagnetic radiation, such as light, occurs in discrete bundles, or "quanta," of energy now known as photons. Bohr postulated that electrons circled the nucleus in orbits of fixed size and energy and that an electron could jump from one orbit to another only by emitting or absorbing specific quanta of energy. By thus incorporating quantization into his theory of the atom, Bohr introduced one of the basic elements of modern particle physics and prompted wider acceptance of quantization to explain atomic and subatomic phenomena.

Size

Subatomic particles play two vital roles in the structure of matter. They are both the basic building blocks of the universe and the mortar that binds the blocks. Although the particles that fulfill these different roles are of two distinct types, they do share some common characteristics, foremost of which is size.

The small size of subatomic particles is perhaps most convincingly expressed not by stating their absolute units of measure but by comparing them with the complex particles of which they are a part. An atom, for instance, is typically 10^{-10} metre across, yet almost all of the size of the atom is unoccupied "empty" space available to the point-charge electrons surrounding the nucleus. The distance across an atomic nucleus of average size is roughly 10^{-14} metre—only $^{1}/_{10,000}$ the diameter of the atom. The nucleus, in turn, is made up of positively charged protons and electrically neutral neutrons, collectively referred to as nucleons, and a single nucleon has a diameter of about 10^{-15} metre—that is, about $^{1}/_{10}$ that of the nucleus and $^{1}/_{10,000}$ that of the atom. (The distance across the nucleon 10^{-15} metre, is known as a fermi, in honour of the Italian-born physicist Enrico Fermi, who did much experimental and theoretical work on the nature of the nucleus and its contents).

The sizes of atoms, nuclei, and nucleons are measured by firing a beam of electrons at an appropriate target. The higher the energy of the electrons, the farther they penetrate before being deflected by the electric charges within the atom. For example, a beam with an energy of a few hundred electron volts (eV) scatters from the electrons in a target atom. The way in which the beam is scattered (electron scattering) can then be studied to determine the general distribution of the atomic electrons.

At energies of a few hundred megaelectron volts ((MeV; 10^{6} eV) electrons in the beam are little affected by atomic electrons; instead, they penetrate the atom and are scattered by the positive nucleus. Therefore, if such a beam is fired at liquid hydrogen, whose atoms contain only single protons in their nuclei, the pattern of scattered electrons reveals the size of the proton. At energies greater than a gigaelectron volt (GeV; 10^{9} eV)), the electrons penetrate within the protons and neutrons, and their scattering patterns reveal an inner structure. Thus, protons and neutrons are no more indivisible than atoms are; indeed, they contain still smaller particles, which are called quarks.

Quarks are as small as or smaller than physicists can measure. In experiments at very high energies, equivalent to probing protons in a target with electrons accelerated to nearly 50,000 GeV, quarks appear to behave as points in space, with no measurable size; they must therefore be smaller than 10^{-18} metre, or less than $^{1}/_{1,000}$ the size of the individual nucleons they form. Similar experiments show that electrons too are smaller than it is possible to measure.

Elementary Particles

Electrons and quarks contain no discernible structure; they cannot be reduced or separated into smaller components. It is therefore reasonable to call them "elementary" particles, a name that in the past was mistakenly given to particles such as the proton, which is in fact a complex particle that contains quarks. The term subatomic particle refers both to the true elementary particles, such as quarks and electrons, and to the larger particles that quarks form.

Although both are elementary particles, electrons and quarks differ in several respects. Whereas quarks together form nucleons within the atomic nucleus, the electrons generally circulate toward the periphery of atoms. Indeed, electrons are regarded as distinct from quarks and are classified in a separate group of elementary particles called leptons. There are several types of lepton, just as there are several types of quark. Only two types of quark are needed to form protons and neutrons, however, and these, together with the electron and one other elementary particle, are all the building blocks that are necessary to build the everyday world. The last particle required is an electrically neutral particle called the neutrino.

Neutrinos do not exist within atoms in the sense that electrons do, but they play a crucial role in certain types of radioactive decay. In a basic process of one type of radioactivity, known as beta decay, a neutron changes into a proton. In making this change, the neutron acquires one unit of positive charge. To keep the overall charge in the beta-decay process constant and thereby conform to the fundamental physical law of charge conservation, the neutron must emit a negatively charged electron. In addition, the neutron also emits a neutrino (strictly speaking, an antineutrino), which has little or no mass and no electric charge. Beta decays are important in the transitions that occur when unstable atomic nuclei change to become more stable, and for this reason neutrinos are a necessary component in establishing the nature of matter.

The neutrino, like the electron, is classified as a lepton. Thus, it seems at first sight that only four kinds of elementary particles—two quarks and two leptons—should exist. In the 1930s, however, long before the concept of quarks was established, it became clear that matter is more complicated.

Spin

The concept of quantization led during the 1920s to the development of quantum mechanics, which appeared to provide physicists with the correct method of calculating the structure of the atom. In his model Niels Bohr had postulated that the electrons in the atom move only in orbits in which the angular momentum (angular velocity multiplied by mass) has certain fixed values. Each of these allowed values is characterized by a quantum number that can have only integer values. In the full quantum mechanical treatment of the structure of the atom, developed in the 1920s, three quantum numbers relating to angular momentum arise because there are three independent variable parameters in the equation describing the motion of atomic electrons.

In 1925, however, two Dutch physicists, Samuel Goudsmit and George Uhlenbeck, realized that, in order to explain fully the spectra of light emitted by the atoms of alkali metals, such as sodium, which have one outer valence electron beyond the main core, there must be a fourth quantum number that can take only two values, $-^1/_2$ and $^1/_2$. Goudsmit and Uhlenbeck proposed that this quantum number refers to an internal angular momentum, or spin, that the electrons possess. This implies that the electrons, in effect, behave like spinning electric charges. Each therefore creates a magnetic field and has its own magnetic moment. The internal magnet of an atomic electron orients itself in one of two directions with respect to the magnetic field created by the rest of the atom. It is either parallel or antiparallel; hence, there are two quantized states—and two possible values of the associated spin quantum number.

The concept of spin is now recognized as an intrinsic property of all subatomic particles. Indeed, spin is one of the key criteria used to classify particles into two main groups: fermions, with half-integer values of spin ($^1/_2$, $^3/_2$,...) and bosons, with integer values of spin (0, 1, 2,...). In the Standard

Model all of the "matter" particles (quarks and leptons) are fermions, whereas "force" particles such as photons are bosons. These two classes of particles have different symmetry properties that affect their behaviour.

Antiparticles

Two years after the work of Goudsmit and Uhlenbeck, the English theorist P.A.M. Dirac provided a sound theoretical background for the concept of electron spin. In order to describe the behaviour of an electron in an electromagnetic field, Dirac introduced the German-born physicist Albert Einstein's theory of special relativity into quantum mechanics. Dirac's relativistic theory showed that the electron must have spin and a magnetic moment, but it also made what seemed a strange prediction. The basic equation describing the allowed energies for an electron would admit two solutions, one positive and one negative. The positive solution apparently described normal electrons. The negative solution was more of a mystery; it seemed to describe electrons with positive rather than negative charge.

The mystery was resolved in 1932, when Carl Anderson, an American physicist, discovered the particle called the positron. Positrons are very much like electrons: they have the same mass and the same spin, but they have opposite electric charge. Positrons, then, are the particles predicted by Dirac's theory, and they were the first of the so-called antiparticles to be discovered. Dirac's theory, in fact, applies to any subatomic particle with spin $^{1}/_{2}$; therefore, all spin$-^{1}/_{2}$ particles should have corresponding antiparticles. Matter cannot be built from both particles and antiparticles, however. When a particle meets its appropriate antiparticle, the two disappear in an act of mutual destruction known as annihilation. Atoms can exist only because there is an excess of electrons, protons, and neutrons in the everyday world, with no corresponding positrons, antiprotons, and antineutrons.

Positrons do occur naturally, however, which is how Anderson discovered their existence. High-energy subatomic particles in the form of cosmic rays continually rain down on Earth's atmosphere from outer space, colliding with atomic nuclei and generating showers of particles that cascade toward the ground. In these showers the enormous energy of the incoming cosmic ray is converted to matter, in accordance with Einstein's theory of special relativity, which states that $E=mc^2$, where E is energy, m is mass, and c is the velocity of light. Among the particles created are pairs of electrons and positrons. The positrons survive for a tiny fraction of a second until they come close enough to electrons to annihilate. The total mass of each electron-positron pair is then converted to energy in the form of gamma-ray photons.

Using particle accelerators, physicists can mimic the action of cosmic rays and create collisions at high energy. In 1955 a team led by the Italian-born scientist Emilio Segrè and the American Owen Chamberlain found the first evidence for the existence of antiprotons in collisions of high-energy protons produced by the Bevatron, an accelerator at what is now the Lawrence Berkeley National Laboratory in California. Shortly afterward, a different team working on the same accelerator discovered the antineutron.

Since the 1960s physicists have discovered that protons and neutrons consist of quarks with spin 1/2 and that antiprotons and antineutrons consist of antiquarks. Neutrinos too have spin 1/2 and therefore have corresponding antiparticles known as antineutrinos. Indeed, it is an antineutrino, rather than a neutrino, that emerges when a neutron changes by beta decay into a proton. This reflects an empirical law regarding the production and decay of quarks and leptons: in any interaction the total numbers of quarks and leptons seem always to remain constant. Thus, the appearance of a lepton—the electron—in the decay of a neutron must be balanced by the simultaneous appearance of an antilepton, in this case the antineutrino.

In addition to such familiar particles as the proton, neutron, and electron, studies have slowly revealed the existence of more than 200 other subatomic particles. These "extra" particles do not appear in the low-energy environment of everyday human experience; they emerge only at the higher energies found in cosmic rays or particle accelerators. Moreover, they immediately decay to the more-familiar particles after brief lifetimes of only fractions of a second. The variety and behaviour of these extra particles initially bewildered scientists but have since come to be understood in terms of the quarks and leptons. In fact, only six quarks, six leptons, and their corresponding antiparticles are necessary to explain the variety and behaviour of all the subatomic particles, including those that form normal atomic matter.

Four Basic Forces

Quarks and leptons are the building blocks of matter, but they require some sort of mortar to bind themselves together into more-complex forms, whether on a nuclear or a universal scale. The particles that provide this mortar are associated with four basic forces that are collectively referred to as the fundamental interactions of matter. These four basic forces are gravity (or the gravitational force), the electromagnetic force, and two forces more familiar to physicists than to laypeople: the strong force and the weak force.

On the largest scales the dominant force is gravity. Gravity governs the aggregation of matter into stars and galaxies and influences the way that the universe has evolved since its origin in the big bang. The best-understood force, however, is the electromagnetic force, which underlies the related phenomena of electricity and magnetism. The electromagnetic force binds negatively charged electrons to positively charged atomic nuclei and gives rise to the bonding between atoms to form matter in bulk.

Gravity and electromagnetism are well known at the macroscopic level. The other two forces act only on subatomic scales, indeed on subnuclear scales. The strong force binds quarks together within protons, neutrons, and other subatomic particles. Rather as the electromagnetic force is ultimately responsible for holding bulk matter together, so the strong force also keeps protons and neutrons together within atomic nuclei. Unlike the strong force, which acts only between quarks, the weak force acts on both quarks and leptons. This force is responsible for the beta decay of a neutron into a proton and for the nuclear reactions that fuel the Sun and other stars.

Field Theory

Since the 1930s physicists have recognized that they can use field theory to describe the interactions of all four basic forces with matter. In mathematical terms a field describes something that varies continuously through space and time. A familiar example is the field that surrounds a piece of magnetized iron. The magnetic field maps the way that the force varies in strength and direction around the magnet. The appropriate fields for the four basic forces appear to have an important property in common: they all exhibit what is known as gauge symmetry. Put simply, this means that certain changes can be made that do not affect the basic structure of the field. It also implies that the relevant physical laws are the same in different regions of space and time.

At a subatomic, quantum level these field theories display a significant feature. They describe each basic force as being in a sense carried by its own subatomic particles. These "force" particles are now called gauge bosons, and they differ from the "matter" particles—the quarks and leptons discussed earlier—in a fundamental way. Bosons are characterized by integer values of their spin quantum number, whereas quarks and leptons have half-integer values of spin.

The most familiar gauge boson is the photon, which transmits the electromagnetic force between electrically charged objects such as electrons and protons. The photon acts as a private, invisible messenger between these particles, influencing their behaviour with the information it conveys, rather as a ball influences the actions of children playing catch. Other gauge bosons, with varying properties, are involved with the other basic forces.

In developing a gauge theory for the weak force in the 1960s, physicists discovered that the best theory, which would always yield sensible answers, must also incorporate the electromagnetic force. The result was what is now called electroweak theory. It was the first workable example of a unified field theory linking forces that manifest themselves differently in the everyday world. Unified theory reveals that the basic forces, though outwardly diverse, are in fact separate facets of a single underlying force. The search for a unified theory of everything, which incorporates all four fundamental forces, is one of the major goals of particle physics. It is leading theorists to an exciting area of study that involves not only subatomic particle physics but also cosmology and astrophysics.

Gravity

The weakest, and yet the most pervasive, of the four basic forces is gravity. It acts on all forms of mass and energy and thus acts on all subatomic particles, including the gauge bosons that carry the forces. The 17th-century English scientist Isaac Newton was the first to develop a quantitative description of the force of gravity. He argued that the force that binds the Moon in orbit around Earth is the same force that makes apples and other objects fall to the ground, and he proposed a universal law of gravitation.

According to Newton's law, all bodies are attracted to each other by a force that depends directly on the mass of each body and inversely on the square of the distance between them. For a pair of masses, m1 and m2, a distance r apart, the strength of the force F is given by:

$$F = Gm_1m_2 / r^2.$$

G is called the constant of gravitation and is equal to 6.67×10^{-11} newton-metre2-kilogram−2.

The constant G gives a measure of the strength of the gravitational force, and its smallness indicates that gravity is weak. Indeed, on the scale of atoms the effects of gravity are negligible compared with the other forces at work. Although the gravitational force is weak, its effects can be extremely long-ranging. Newton's law shows that at some distance the gravitational force between two bodies becomes negligible but that this distance depends on the masses involved. Thus, the gravitational effects of large, massive objects can be considerable, even at distances far outside the range of the other forces. The gravitational force of Earth, for example, keeps the Moon in orbit some 384,400 km (238,900 miles) distant. Newton's theory of gravity proves adequate for many applications. In 1915, however, the German-born physicist Albert Einstein developed the theory of general relativity, which incorporates the concept of gauge symmetry and yields subtle corrections to Newtonian gravity. Despite its importance, Einstein's general relativity remains a classical theory in the sense that it does not incorporate the ideas of quantum mechanics. In a quantum theory of gravity, the gravitational force must be carried by a suitable messenger particle, or gauge boson. No workable quantum theory of gravity has yet been developed, but general relativity determines some of the properties of the hypothesized "force" particle of gravity, the so-called graviton. In particular, the graviton must have a spin quantum number of 2 and no mass, only energy.

Electromagnetism

The first proper understanding of the electromagnetic force dates to the 18th century, when a French physicist, Charles Coulomb, showed that the electrostatic force between electrically charged objects follows a law similar to Newton's law of gravitation. According to Coulomb's law, the force F between one charge, q_1,, and a second charge, q_2 is proportional to the product of the charges divided by the square of the distance r between them, or $F = kq_1q_2 / r^2$. Here k is the proportionality constant, equal to ${}^1/_4\pi\varepsilon_0$ (ε_0 (εo being the permittivity of free space). An electrostatic force can be either attractive or repulsive, because the source of the force, electric charge, exists in opposite forms: positive and negative. The force between opposite charges is attractive, whereas bodies with the same kind of charge experience a repulsive force. Coulomb also showed that the force between magnetized bodies varies inversely as the square of the distance between them. Again, the force can be attractive (opposite poles) or repulsive (like poles).

Magnetism and electricity are not separate phenomena; they are the related manifestations of an underlying electromagnetic force. Experiments in the early 19th century by, among others, Hans Ørsted (in Denmark), André-Marie Ampère (in France), and Michael Faraday (in England) revealed the intimate connection between electricity and magnetism and the way the one can give rise to the other. The results of these experiments were synthesized in the 1850s by the Scottish physicist James Clerk Maxwell in his electromagnetic theory. Maxwell's theory predicted the existence of electromagnetic waves—undulations in intertwined electric and magnetic fields, traveling with the velocity of light Max Planck's work in Germany at the turn of the 20th century, in which he explained the spectrum of radiation from a perfect emitter (blackbody radiation), led to the concept of quantization and photons. In the quantum picture, electromagnetic radiation has a dual nature, existing both as Maxwell's waves and as streams of particles called photons. The quantum nature of electromagnetic radiation is encapsulated in quantum electrodynamics, the quantum field theory of the electromagnetic force. Both Maxwell's classical theory and the quantized version contain gauge symmetry, which now appears to be a basic feature of the fundamental forces.

The electromagnetic force is intrinsically much stronger than the gravitational force. If the relative strength of the electromagnetic force between two protons separated by the distance within the nucleus was set equal to one, the strength of the gravitational force would be only 10^{-36}. At an atomic level the electromagnetic force is almost completely in control; gravity dominates on a large scale only because matter as a whole is electrically neutral.

The gauge boson of electromagnetism is the photon, which has zero mass and a spin quantum number of 1. Photons are exchanged whenever electrically charged subatomic particles interact. The photon has no electric charge, so it does not experience the electromagnetic force itself; in other words, photons cannot interact directly with one another. Photons do carry energy and momentum, however, and, in transmitting these properties between particles, they produce the effects known as electromagnetism.

In these processes energy and momentum are conserved overall (that is, the totals remain the same, in accordance with the basic laws of physics), but, at the instant one particle emits a photon and another particle absorbs it, energy is not conserved. Quantum mechanics allows this imbalance, provided that the photon fulfills the conditions of Heisenberg's uncertainty principle. This rule, described in 1927 by the German scientist Werner Heisenberg, states that it is impossible, even in principle, to know all the details about a particular quantum system. For example, if the exact position of an electron is identified, it is impossible to be certain of the electron's momentum. This fundamental uncertainty allows a discrepancy in energy, ΔE, to exist for a time, Δt, provided that the product of ΔE and Δt is very small—equal to the value of Planck's constant divided by 2π, *or* 1.05×10^{-34} joule seconds. The energy of the exchanged photon can thus be thought of as "borrowed," within the limits of the uncertainty principle (i.e., the more energy borrowed, the shorter the time of the loan). Such borrowed photons are called "virtual" photons to distinguish them from real photons, which constitute electromagnetic radiation and can, in principle, exist forever. This concept of virtual particles in processes that fulfill the conditions of the uncertainty principle applies to the exchange of other gauge bosons as well.

Weak Force

Since the 1930s physicists have been aware of a force within the atomic nucleus that is responsible for certain types of radioactivity that are classed together as beta decay. A typical example of beta decay occurs when a neutron transmutes into a proton. The force that underlies this process is known as the weak force to distinguish it from the strong force that binds quarks together.

The correct gauge field theory for the weak force incorporates the quantum field theory of electromagnetism (quantum electrodynamics) and is called electroweak theory. It treats the weak force and the electromagnetic force on an equal footing by regarding them as different manifestations of a more-fundamental electroweak force, rather as electricity and magnetism appear as different aspects of the electromagnetic force.

The electroweak theory requires four gauge bosons. One of these is the photon of electromagnetism; the other three are involved in reactions that occur via the weak force. These weak gauge bosons include two electrically charged versions, called W+ and W−, where the signs indicate the charge, and a neutral variety called Zo, where the zero indicates no charge. Like the photon, the W and Z particles have a spin quantum number of 1; unlike the photon, they are very massive. The W particles have a mass of about 80.4 GeV, while the mass of the Zo particle is 91.187 GeV. By comparison, the mass of the proton is 0.94 GeV, or about one-hundredth that of the Z particle. (Strictly speaking, mass should be given in units of energy/c2, where c is the velocity of light. However, common practice is to set c = 1 so that mass is quoted simply in units of energy, eV, as in this paragraph).

The charged W particles are responsible for processes, such as beta decay, in which the charge of the participating particles changes hands. For example, when a neutron transmutes into a proton, it emits a W−; thus, the overall charge remains zero before and after the decay process. The W particle involved in this process is a virtual particle. Because its mass is far greater than that of the neutron, the only way that it can be emitted by the lightweight neutron is for its existence to be fleetingly short, within the requirements of the uncertainty principle. Indeed, the W− immediately transforms into an electron and an antineutrino, the particles that are observed in the laboratory as the products of neutron beta decay. Z particles are exchanged in similar reactions that involve no change in charge.

In the everyday world the weak force is weaker than the electromagnetic force but stronger than the gravitational force. Its range, however, is very short. Because of the large amounts of energy needed to create the large masses of the W and Z particles, the uncertainty principle ensures that a weak gauge boson cannot be borrowed for long, which limits the range of the force to distances less than 10−17 metre. The weak force between two protons in a nucleus is only 10−7 the strength of the electromagnetic force. As the electroweak theory reveals and as experiments confirm, however, this weak force becomes effectively stronger as the energies of the participating particles increase. When the energies reach 100 GeV or so—roughly the energy equivalent to the mass of the W and Z particles—the strength of the weak force becomes comparable to that of the electromagnetic force. This means that reactions that involve the exchange of a Z0 become as common as those in which a photon is exchanged. Moreover, at these energies real W and Z particles, as opposed to virtual ones, can be created in reactions.

Unlike the photon, which is stable and can in principle live forever, the heavy weak gauge bosons decay to lighter particles within an extremely brief lifetime of about 10−25 second. This is roughly a million million times shorter than experiments can measure directly, but physicists can detect the particles into which the W and Z particles decay and can thus infer their existence.

The Strong Force

Although the aptly named strong force is the strongest of all the fundamental interactions, it, like the weak force, is short-ranged and is ineffective much beyond nuclear distances of 10^{-15} metre or so. Within the nucleus and, more specifically, within the protons and other particles that are built from quarks, however, the strong force rules supreme; between quarks in a proton, it can be almost 100 times stronger than the electromagnetic force, depending on the distance between the quarks.

During the 1970s physicists developed a theory for the strong force that is similar in structure to quantum electrodynamics. In this theory quarks are bound together within protons and neutrons by exchanging gauge bosons called gluons. The quarks carry a property called "colour" that is analogous to electric charge. Just as electrically charged particles experience the electromagnetic force and exchange photons, so colour-charged, or coloured, particles feel the strong force and exchange gluons. This property of colour gives rise in part to the name of the theory of the strong force: quantum chromodynamics.

Gluons are massless and have a spin quantum number of 1. In this respect they are much like photons, but they differ from photons in one crucial way. Whereas photons do not interact among themselves—because they are not electrically charged—gluons do carry colour charge. This means that gluons can interact together, which has an important effect in limiting the range of gluons and in confining quarks within protons and other particles.

There are three types of colour charge, called red, green, and blue, although there is no connection between the colour charge of quarks and gluons and colour in the usual sense. Quarks each carry a single colour charge, while gluons carry both a colour and an anticolour charge.

The strong force acts in such a way that quarks of different colour are attracted to one another; thus, red attracts green, blue attracts red, and so on. Quarks of the same colour, on the other hand, repel each other. The quarks can combine only in ways that give a net colour charge of zero. In particles that contain three quarks, such as protons, this is achieved by adding red, blue, and green. An alternative, observed in particles called mesons, is for a quark to couple with an antiquark of the same basic colour. In this case the colour of the quark and the anticolour of the antiquark cancel each other out. These combinations of three quarks (or three antiquarks) or of quark-antiquark pairs are the only combinations that the strong force seems to allow.

The constraint that only colourless objects can appear in nature seems to limit attempts to observe single quarks and free gluons. Although a quark can radiate a real gluon just as an electron can radiate a real photon, the gluon never emerges on its own into the surrounding environment. Instead, it somehow creates additional gluons, quarks, and antiquarks from its own energy and materializes as normal particles built from quarks. Similarly, it appears that the strong force keeps quarks permanently confined within larger particles. Attempts to knock quarks out of protons by, for example, knocking protons together at high energies succeed only in creating more particles—that is, in releasing new quarks and antiquarks that are bound together and are themselves confined by the strong force.

Classes of Subatomic Particles

From the early 1930s to the mid-1960s, studies of the composition of cosmic rays and experiments using particle accelerators revealed more than 200 types of subatomic particles. In order to comprehend this rich variety, physicists began to classify the particles according to their properties (such as mass, charge, and spin) and to their behaviour in response to the fundamental interactions—in particular, the weak and strong forces. The aim was to discover common features that would simplify the variety, much as the periodic table of chemical elements had done for the wealth of atoms discovered in the 19th century. An important result was that many of the particles, those classified as hadrons, were found to be composed of a much smaller number of more-elementary particles, the quarks. Today the quarks, together with the group of leptons, are recognized as fundamental particles of matter.

Leptons and Antileptons

Leptons are a group of subatomic particles that do not experience the strong force. They do, however, feel the weak force and the gravitational force, and electrically charged leptons interact via the electromagnetic force. In essence, there are three types of electrically charged leptons and three types of neutral leptons, together with six related antileptons. In all three cases the charged lepton has a negative charge, whereas its antiparticle is positively charged. Physicists coined the name lepton from the Greek word for "slender" because, before the discovery of the tau in 1975, it seemed that the leptons were the lightest particles. Although the name is no longer appropriate, it has been retained to describe all spin-1/2 particles that do not feel the strong force.

Charged Leptons (Electron, Muon, Tau)

Probably the most-familiar subatomic particle is the electron, the component of atoms that makes interatomic bonding and chemical reactions—and hence life—possible. The electron was also the first particle to be discovered. Its negative charge of 1.6 × 10−19 coulomb seems to be the basic unit of electric charge, although theorists have a poor understanding of what determines this particular size.

The electron, with a mass of 0.511 megaelectron volts (MeV; 106 eV), is the lightest of the charged leptons. The next-heavier charged lepton is the muon. It has a mass of 106 MeV, which is some 200 times greater than the electron's mass but is significantly less than the proton's mass of 938 MeV. Unlike the electron, which appears to be completely stable, the muon decays after an average lifetime of 2.2 millionths of a second into an electron, a neutrino, and an antineutrino. This process, like the beta decay of a neutron into a proton, an electron, and an antineutrino, occurs via the weak force. Experiments have shown that the intrinsic strength of the underlying reaction is the same in both kinds of decay, thus revealing that the weak force acts equally upon leptons (electrons, muons, neutrinos) and quarks (which form neutrons and protons).

There is a third, heavier type of charged lepton, called the tau. The tau, with a mass of 1,777 MeV, is even heavier than the proton and has a very short lifetime of about 10−13 second. Like the electron and the muon, the tau has its associated neutrino. The tau can decay into a muon, plus a tau-neutrino and a muon-antineutrino; or it can decay directly into an electron, plus a tau-neutrino and an electron-antineutrino. Because the tau is heavy, it can also decay into particles containing quarks. In one example the tau decays into particles called pi-mesons, which are accompanied by a tau-neutrino.

Neutral Leptons (Neutrino)

Unlike the charged leptons, the electrically neutral leptons, the neutrinos, do not come under the influence of the electromagnetic force. They experience only the weakest two of nature's forces, the weak force and gravity. For this reason neutrinos react extremely weakly with matter. They can, for example, pass through Earth without interacting, which makes it difficult to detect neutrinos and to measure their properties.

Although electrically neutral, the neutrinos seem to carry an identifying property that associates them specifically with one type of charged lepton. In the example of the muon's decay, the antineutrino produced is not simply the antiparticle of the neutrino that appears with it. The neutrino carries a muon-type hallmark, while the antineutrino, like the antineutrino emitted when a neutron decays, is always an electron-antineutrino. In interactions with matter, such electron-neutrinos and antineutrinos never produce muons, only electrons. Likewise, muon-neutrinos give rise to muons only, never to electrons.

Theory does not require the mass of neutrinos to be any specific amount, and in the past it was assumed to be zero. Observations of the cosmic microwave background and gravitational lensing of distant galaxies indicate the three flavours of neutrino have a combined mass of 0.32 eV, or less than $^{1}/_{100,000,000}$ the mass of an electron. Neutrinos can change from one type to another, or "oscillate." This can happen only if the neutrino types in question have small differences in mass—and hence must have mass.

Hadrons

The name hadron comes from the Greek word for "strong"; it refers to all those particles that are built from quarks and therefore experience the strong force. The most common examples of this class are the proton and the neutron, the two types of particle that build up the nucleus of every atom.

Stable and Resonant Hadrons

Experiments have revealed a large number of hadrons, of which only the proton appears to be stable. Indeed, even if the proton is not absolutely stable, experiments show that its lifetime is at least in excess of 5.9×10^{33} years. In contrast, a single neutron, free from the forces at work within the nucleus, lives an average of nearly 15 minutes before decaying. Within a nucleus, however—even the simple nucleus of deuterium, which consists of one proton and one neutron—the balance of forces is sufficient to prolong the neutron's lifetime so that many nuclei are stable and a large variety of chemical elements exist.

Some hadrons typically exist only 10^{-10} to 10^{-8} second. Fortunately for experimentalists, these particles are usually born in such high-energy collisions that they are moving at velocities close to the speed of light. Their timescale is therefore "stretched" or "slowed down" so that, in the high-speed particle's frame of reference, its lifetime may be 10^{-10} second, but, in a stationary observer's frame of reference, the particle lives much longer. This effect, known as time dilation in the theory of special relativity, allows stationary particle detectors to record the tracks left by these short-lived particles. These hadrons, which number about a dozen, are usually referred to as "stable" to distinguish them from still shorter-lived hadrons with lifetimes typically in the region of a mere 10^{-23} second.

The stable hadrons usually decay via the weak force. In some cases they decay by the electromagnetic force, which results in somewhat shorter lifetimes because the electromagnetic force is stronger than the weak force. The very-short-lived hadrons, however, which number 200 or more, decay via the strong force. This force is so strong that it allows the particles to live only for about the time it takes light to cross the particle; the particles decay almost as soon as they are created.

These very-short-lived particles are called "resonant" because they are observed as a resonance phenomenon; they are too short-lived to be observed in any other way. Resonance occurs when a system absorbs more energy than usual because the energy is being supplied at the system's own natural frequency. For example, soldiers break step when they cross a bridge because their rhythmic marching could make the bridge resonate—set it vibrating at its own natural frequency—so that it absorbs enough energy to cause damage. Subatomic-particle resonances occur when the net energy of colliding particles is just sufficient to create the rest mass of the new particle, which the strong force then breaks apart within 10^{-23} second. The absorption of energy, or its subsequent emission in the form of particles as the resonance decays, is revealed as the energy of the colliding particles is varied.

Baryons and Mesons

The hadrons, whether stable or resonant, fall into two classes: baryons and mesons. Originally the names referred to the relative masses of the two groups of particles. The baryons included the proton and heavier particles; the mesons were particles with masses between those of the electron and the proton. Now, however, the name baryon refers to any particle built from three quarks, such as the proton and the neutron. Mesons, on the other hand, are particles built from a quark combined with an antiquark. As described in the section The strong force, these are the only two combinations of quarks and antiquarks that the strong force apparently allows.

The two groups of hadrons are also distinguished from one another in terms of a property called baryon number. The baryons are characterized by a baryon number, B, of 1; antibaryons have a baryon number of −1; and the baryon number of the mesons, leptons, and messenger particles is 0. Baryon numbers are additive; thus, an atom containing one proton and one neutron (each with a baryon number of 1) has a baryon number of 2. Quarks therefore must have a baryon number of 1/3, and the antiquarks a baryon number of −1/3, in order to give the correct values of 1 or 0 when they combine to form baryons and mesons.

The empirical law of baryon conservation states that in any reaction the total number of baryons must remain constant. If any baryons are created, then so must be an equal number of antibaryons, which in principle negate the baryons. Conservation of baryon number explains the apparent stability of the proton. The proton does not decay into lighter positive particles, such as the positron or the mesons, because those particles have a baryon number of 0. Neutrons and other heavy baryons can decay into the lighter protons, however, because the total number of baryons present does not change.

At a more-detailed level, baryons and mesons are differentiated from one another in terms of their spin. The basic quarks and antiquarks have a spin of 1/2 (which may be oriented in either of two directions). When three quarks combine to form a baryon, their spins can add up to only half-integer values. In contrast, when quarks and antiquarks combine to form mesons, their spins always add up to integer values. As a result, baryons are classified as fermions within the Standard Model of particle physics, whereas mesons are classified as bosons.

Quarks and Antiquarks

The baryons and mesons are complex subatomic particles built from more-elementary objects, the quarks. Six types of quark, together with their corresponding antiquarks, are necessary to account for all the known hadrons. The six varieties, or "flavours," of quark have acquired the names up, down, charm, strange, top, and bottom. The meaning of these somewhat unusual names is not important; they have arisen for a number of reasons. What is important is the way that the quarks contribute to matter at different levels and the properties that they bear.

The quarks are unusual in that they carry electric charges that are smaller in magnitude than e, the size of the charge of the electron (1.6×10^{-19} coulomb). This is necessary if quarks are to combine together to give the correct electric charges for the observed particles, usually 0, +e, or −e. Only two types of quark are necessary to build protons and neutrons, the constituents of atomic nuclei. These are the up quark, with a charge of + $+{}^{2}/_{3}\,e$, and the down quark, which has a charge of – $^{-}{}^{1}/_{3}\,e$. The proton consists of two up quarks and one down quark, which gives it a total charge of $+e$.

The neutron, on the other hand, is built from one up quark and two down quarks, so that it has a net charge of zero. The other properties of the up and down quarks also add together to give the measured values for the proton and neutron. For example, the quarks have spins of $^{1}/_{2}$. In order to form a proton or a neutron, which also have spin $^1/_2$ the quarks must align in such a way that two of the three spins cancel each other, leaving a net value of $^1/_2$.

Up and down quarks can also combine to form particles other than protons and neutrons. For example, the spins of the three quarks can be arranged so that they do not cancel. In this case they form short-lived resonance states, which have been given the name delta, or Δ. The deltas have spins of $^3/_2$ and the up and down quarks combine in four possible configurations—uuu, uud, udd, and ddd—where u and d stand for up and down. The charges of these Δ states are $+2e,\ +e,\ 0,\ and\ -e$, respectively.

The up and down quarks can also combine with their antiquarks to form mesons. The pi-meson, or pion, which is the lightest meson and an important component of cosmic rays, exists in three forms: with charge e (or 1), with charge 0, and with charge −e (or −1). In the positive state an up quark combines with a down antiquark; a down quark together with an up antiquark compose the negative pion; and the neutral pion is a quantum mechanical mixture of two states—$u\overline{u}$ and $d\overline{d}$, where the bar over the top of the letter indicates the antiquark.

Up and down are the lightest varieties of quarks. Somewhat heavier are a second pair of quarks, charm (c) and strange (s), with charges of $+\frac{2}{3}e$ and $-\frac{1}{3}e$, respectively. A third, still heavier pair of quarks consists of top (or truth, t) and bottom (or beauty, b), again with charges of $+\frac{2}{3}e$ and $-\frac{1}{3}e$ respectively. These heavier quarks and their antiquarks combine with up and down quarks and with each other to produce a range of hadrons, each of which is heavier than the basic proton and pion, which represent the lightest varieties of baryon and meson, respectively. For example, the particle called lambda (Λ) is a baryon built from u, d, and s quarks; thus, it is like the neutron but with a d quark replaced by an s quark.

Fermions

In particle physics, a fermion is a type of particle that obeys the rules of Fermi-Dirac statistics, namely the Pauli Exclusion Principle. These fermions also have a quantum spin with contains a half-integer value, such as 1/2, -1/2, -3/2, and so on. (By comparison, there are other types of particles, called bosons, that have an integer spin, such as 0, 1, -1, -2, 2, etc).

Fermions are sometimes called matter particles, because they are the particles that make up most of what we think of as physical matter in our world, including protons, neutrons, and electrons.

Fermions were first predicted in 1925 by the physicist Wolfgang Pauli, who was trying to figure out how to explain the atomic structure proposed in 1922 by Niels Bohr. Bohr had used experimental evidence to build an atomic model which contained electron shells, creating stable orbits for electrons to move around the atomic nucleus. Though this matched well with the evidence, there was no particular reason why this structure would be stable and that's the explanation that Pauli was trying to reach. He realized that if you assigned quantum numbers (later named quantum spin) to these electrons, then there seemed to be some sort of principle which meant that no two of the electrons could be in exactly the same state. This rule became known as the Pauli Exclusion Principle.

In 1926, Enrico Fermi and Paul Dirac independently tried to understand other aspects of seemingly-contradictory electron behavior and, in doing so, established a more complete statistical way of dealing with electrons. Though Fermi developed the system first, they were close enough and both did enough work that posterity has dubbed their statistical method Fermi-Dirac statistics, though the particles themselves were named after Fermi himself.

The fact that fermions cannot all collapse into the same state - again, that's the ultimate meaning of the Pauli Exclusion Principle - is very important. The fermions within the sun (and all other stars) are collapsing together under the intense force of gravity, but they cannot fully collapse because of the Pauli Exclusion Principle. As a result, there is a pressure generated that pushes against the gravitational collapse of the star's matter. It is this pressure which generates the solar heat that fuels not only our planet but so much of the energy in the rest of our universe, including the very formation of heavy elements, as described by stellar nucleosynthesis.

Fundamental Fermions

There are a total of 12 fundamental fermions - fermions that aren't made up of smaller particles - that have been experimentally identified. They fall into two categories:

- Quarks - Quarks are the particles that make up hadrons, such as protons and neutrons. There are 6 distinct types of quarks:
    - Up Quark,
    - Charm Quark,
    - Top Quark,
    - Down Quark,
    - Strange Quark,
    - Bottom Quark.
- Leptons - There are 6 types of leptons:
    - Electron,
    - Electron Neutrino,
    - Muon,
    - Muon Neutrino,
    - Tau,
    - Tau Neutrino.

In addition to these particles, the theory of supersymmetry predicts that every boson would have a so-far-undetected fermionic counterpart. Since there are 4 to 6 fundamental bosons, this would suggest that - if supersymmetry is true - there are another 4 to 6 fundamental fermions that have not yet been detected, presumably because they are highly unstable and have decayed into other forms.

Composite Fermions

Beyond the fundamental fermions, another class of fermions can be created by combining fermions together (possibly along with bosons) to get a resulting particle with a half-integer spin. The quantum spins add up, so some basic mathematics shows that any particle which contains an odd number of fermions is going to end up with a half-integer spin and, therefore, will be a fermion itself. Some examples include:

- Baryons - These are particles, like protons and neutrons, that are composed of three quarks joined together. Since each quark has a half-integer spin, the resulting baryon will always have a half-integer spin, no matter which three types of quark join together to form it.
- Helium-3 - Contains 2 protons and 1 neutron in the nucleus, along with 2 electrons circling it. Since there is an odd number of fermions, the resulting spin is a half-integer value. This means that helium-3 is a fermion as well.

Quarks

In particle physics, a quark is one of the elementary (or fundamental) particles that are the building blocks of matter. Elementary particles are classified as fermions and bosons, and fermions are subdivided into quarks and leptons. Quarks are fermions that experience the strong interaction (or strong nuclear force), which involves coupling with the bosons known as gluons. In other words, quarks couple with gluons to form composite particles such as protons and neutrons. By comparison, a lepton is a fermion that does not experience the strong interaction and does not couple with gluons.

Leptons and quarks come in pairs, and in three generations. Everyday matter is composed of the first generation: Two leptons, the electron and electron-neutrino; and two quarks, called Up and Down.

As is the case for all fundamental particles, the quark is a unified entity of wave and particle, which is known as the “wave-particle duality” of quantum physics. The particle aspect of the quark is point-like even at scales thousands of times smaller than the proton size. The wave aspect of the quark extends over the size of the atomic nucleus. The usual convention is to refer to such unified wave-particle fundamental entities as just “particles.”

Quantum Spin and Probability

All particles (fundamental and composite) can be placed in one of two classes, distinguished by their quantum spin and the type of quantum probability statistics they obey: Fermi-Dirac probability or Bose-Einstein probability, neither of which is like classical probability. (A rough illustration of the difference is that the probability of two classical coins coming up the same is 50 percent, while for two fermion coins it is 0 percent and for two boson coins it is 100 percent).

Both the quark and the electron are fermions with quantum spin -½, giving them the odd property of having to be rotated 720° in order to get back to where you started. (A familiar example of this sort of behavior is the Moebius Strip.) As far as everyday matter is concerned, these two types of fermions are essentially "eternal" and can be considered the "pixels of matter" out of which the physical world is constructed. The photon and gluon are bosons with quantum spin -1; they take only the usual 360° to return to the start. The bosons are ephemeral and "couple" the fundamental interactions of the fermions; they can be considered the "pixels of force" that hold all the fermions together as matter.

It is thought that during the first moments of Creation the temperature was so high that quarks could fly free, just like the electron can today. However, in all conditions found in the current universe—even in supernovae—there are no isolated, free quarks; they are confined by their color charge into colorless combinations of pairs or triplets. All such combinations of quarks are given the generic term hadron. The electron, by contrast, is a lepton.

The quark hadrons are further subdivided into two classes. There are the fermion "pixels of matter" composed of three quarks, the baryons such as the protons and neutrons. Then there are the boson "pixels of force" composed of a quark-antiquark pair, the mesons such as the pions that bind the atomic nucleus together.

The fundamental fermions come in three generations. (The bosons do not.) The quarks and electrons that make up regular matter are all members of the first generation. In this generation, there are two "flavors" of quark, the U- and D-quark (or Up and Down quarks), and two flavors of lepton, the electron and the neutrino. A proton is composed of one D- and two U-quarks; the neutron is one U- and two D-quarks.

In the second generation of fundamental fermions, the pattern is repeated, the only difference being that the particles are much more massive that their first generation counterparts; otherwise they are identical. There are the two quarks, the S- and C-quarks (or Strange and Charm), and the two leptons, the muon and muon-neutrino. The third, and apparently final, generation has the B- and T-quarks (or Bottom and Top) with the tau and the tau-neutrino. These are much more massive than the second generation, but otherwise identical. While abundant in the first moments of Creation, the second and third generations play no apparent role in the current universe, which prompted the famous "Who ordered that?" exclamation by theorist Isidor I. Rabi when the muon was first identified in cosmic ray showers.

Quarks are the only fundamental particles that interact through all four of the fundamental forces. Ignoring gravity, quarks can couple with—create and absorb—the gluons of the strong force, the photons of the electromagnetic force, and the vector bosons of the weak force. In contrast, the electron can couple with photons and vector bosons, while the neutrino can only couple with vector bosons.

The color charge on the quarks comes in three paired varieties (unlike the single positive-negative pair of the electric charge) called red-antired, blue-antiblue, and green-antigreen. The colorless baryons with three quarks have one each of R, G, and B. Rather like the pixels of an RGB TV, all three together make white which accounts for the terminology. It must be emphasized, however, that the color charges on the quarks have nothing to do with the colors of everyday life.

Free Quarks

No search for free quarks or fractional electric charges has returned convincing evidence. The absence of free quarks has therefore been incorporated into the notion of confinement, which, it is believed, the theory of quarks must possess.

Confinement began as an experimental observation, and is expected to follow from the modern theory of strong interactions, called quantum chromodynamics (QCD). Although there is no mathematical derivation of confinement in QCD, it is easy to show using lattice gauge theory.

However, it may be possible to change the confinement by creating dense or hot quark matter. These new phases of QCD matter have been predicted theoretically, and experimental searches for them have now started.

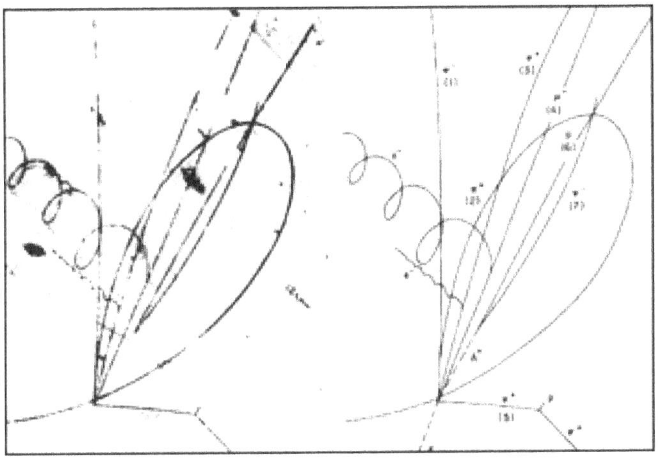

1974 discovery photograph of a possible charmed baryon, now identified as the Σc^{++}.

Confinement and Quark Properties

Every subatomic particle is completely described by a small set of observables such as mass m and quantum numbers, such as spin S and parity P. Usually these properties are directly determined by experiments. However, confinement makes it impossible to measure these properties of quarks. Instead, they must be inferred from measurable properties of the composite particles which are made up of quarks. Such inferences are usually most easily made for certain additive quantum numbers called flavors.

The composite particles made of quarks and antiquarks are the hadrons. These include the mesons which get their quantum numbers from a quark and an antiquark, and the baryons, which get theirs from three quarks. The quarks (and antiquarks) that impart quantum numbers to hadrons are called valence quarks. Apart from these, any hadron may contain an indefinite number of virtual quarks, antiquarks, and gluons which together contribute nothing to their quantum numbers. Such virtual quarks are called sea quarks.

Flavor

Each quark is assigned a baryon number, B = 1/3, and a vanishing lepton number, L = 0. They have fractional electric charge, Q, either Q = +2/3 or Q = −1/3. The former are called up-type quarks, the latter, down-type quarks. Each quark is assigned a weak isospin: Tz = +1/2 for an up-type quark and T_z = −1/2 for a down-type quark. Each doublet of weak isospin defines a generation of quarks. There are three generations, and hence six flavors of quarks—the up-type quark flavors are up, charm, and top; the down-type quark flavors are down, strange, and bottom (each list is in the order of increasing mass).

The number of generations of quarks and leptons are equal in the standard model. The number of generations of leptons with a light neutrino is strongly constrained by experiments at the LEP in CERN and by observations of the abundance of helium in the universe. Precision measurement of the lifetime of the Z boson at LEP constrains the number of light neutrino generations to be three. Astronomical observations of helium abundance give consistent results. Results of direct searches for a fourth generation give limits on the mass of the lightest possible fourth-generation quark. The most stringent limit comes from analysis of results from the Tevatron collider at Fermilab, and shows that the mass of a fourth-generation quark must be greater than 190 GeV. Additional limits on extra quark generations come from measurements of quark mixing performed by the experiments Belle and BaBar.

Each flavor defines a quantum number which is conserved under the strong interactions, but not the weak interactions. The magnitude of flavor changing in the weak interaction is encoded into a structure called the CKM matrix. This also encodes the CP violation allowed in the Standard Model.

Spin

Quantum numbers corresponding to non-Abelian symmetries like rotations require more care in extraction, since they are not additive. In the quark model one builds mesons out of a quark and an antiquark, whereas baryons are built from three quarks. Since mesons are bosons (having integer spins) and baryons are fermions (having half-integer spins), the quark model implies that quarks are fermions. Further, the fact that the lightest baryons have spin-1/2 implies that each quark can have spin S = 1/2. The spins of excited mesons and baryons are completely consistent with this assignment.

Color

Since quarks are fermions, the Pauli exclusion principle implies that the three valence quarks must be in an antisymmetric combination in a baryon. However, the charge Q = 2 baryon Δ^{++} (which is one of four isospin I_z = 3/2 baryons), can only be made of three u quarks with parallel spins. Since this configuration is symmetric under interchange of the quarks, it implies that there exists another internal quantum number, which would then make the combination antisymmetric. This is given the name "color," although it has nothing to do with the perception of the frequency (or wavelength) of light, which is the usual meaning of color. This quantum number is the charge involved in the gauge theory called quantum chromodynamics (QCD).

The only other colored particle is the gluon, which is the gauge boson of QCD. Like all other non-Abelian gauge theories (and unlike quantum electrodynamics), the gauge bosons interact with one another by the same force that affects the quarks.

Color is a gauged SU(3) symmetry. Quarks are placed in the fundamental representation and hence come in three colors (red, green, and blue). Gluons are placed in the adjoint representation and hence come in eight varieties.

Quark Masses

Although one speaks of quark mass in the same way as the mass of any other particle, the notion of mass for quarks is complicated by the fact that quarks cannot be found free in nature. As a result, the notion of a quark mass is a theoretical construct, which makes sense only when one specifies exactly the procedure used to define it.

Current Quark Mass

The approximate chiral symmetry of quantum chromodynamics, for example, allows one to define the ratio between various (up, down, and strange) quark masses through combinations of the masses of the pseudo-scalar meson octet in the quark model through chiral perturbation theory, giving:

$$\frac{m_u}{m_d} = 0.56 \quad \text{and} \quad \frac{m_s}{m_d} = 20.1.$$

The fact that the up quark has mass is important, since there would be no strong CP problem if it were massless. The absolute values of the masses are currently determined from QCD sum rules (also called spectral function sum rules) and lattice QCD. Masses determined in this manner are called current quark masses. The connection between different definitions of the current quark masses needs the full machinery of renormalization for its specification.

Valence Quark Mass

Another, older, method of specifying the quark masses was to use the Gell-Mann-Nishijima mass formula in the quark model, which connect hadron masses to quark masses. The masses so determined are called constituent quark masses, and are significantly different from the current quark masses defined above. The constituent masses do not have any further dynamical meaning.

Heavy Quark Masses

The masses of the heavy charm and bottom quarks are obtained from the masses of hadrons containing a single heavy quark (and one light antiquark or two light quarks) and from the analysis of quarkonia. Lattice QCD computations using the heavy quark effective theory (HQET) or non-relativistic quantum chromodynamics (NRQCD) are currently used to determine these quark masses.

The top quark is sufficiently heavy that perturbative QCD can be used to determine its mass. Before its discovery in 1995, the best theoretical estimates of the top quark mass were obtained from global analysis of precision tests of the Standard Model. The top quark, however, is unique among quarks in that it decays before having a chance to hadronize. Thus, its mass can be directly measured from the resulting decay products. This can only be done at the Tevatron which is the only particle accelerator energetic enough to produce top quarks in abundance.

Properties of Quarks

Generation	Weak Isospin	Flavor	Name	Symbol	Charge / e	Mass / MeV·c⁻²	Antiparticle	Symbol
1	+½	I_z=+½	Up	u	+⅔	1.5 – 4.0	Antiup	$\overline{u}$
1	-½	I_z=-½	Down	d	-⅓	4 – 8	Antidown	$\overline{d}$
2	-½	S=-1	Strange	s	-⅓	80 – 130	Antistrange	$\overline{s}$
2	+½	C=1	Charm	c	+⅔	1150 – 1350	Anticharm	$\overline{c}$
3	-½	B'=-1	Bottom	b	-⅓	4100 – 4400	Antibottom	$\overline{b}$
3	+½	T=1	Top	t	+⅔	170900 ± 1800	Antitop	$\overline{t}$

The following table summarizes the key properties of the six known quarks:

- Top quark mass from Tevatron Electroweak Working Group.
- Other quark masses from Particle Data Group; these masses are given in the MS-bar scheme.
- The quantum numbers of the top and bottom quarks are sometimes known as truth and beauty respectively, as an alternative to topness and bottomness.

Antiquarks

The additive quantum numbers of antiquarks are equal in magnitude and opposite in sign to those of the quarks. CPT symmetry forces them to have the same spin and mass as the corresponding quark. Tests of CPT symmetry cannot be performed directly on quarks and antiquarks, due to confinement, but can be performed on hadrons. Notation of antiquarks follows that of antimatter in general: An up quark is denoted by ū, and an anti-up quark is denoted by ū.

Substructure

Some extensions of the Standard Model begin with the assumption that quarks and leptons have substructure. In other words, these models assume that the elementary particles of the Standard Model are in fact composite particles, made of some other elementary constituents. Such an assumption is open to experimental tests, and these theories are severely constrained by data. At present there is no evidence for such substructure.

Leptons

In particle physics, a lepton is one of the elementary (or fundamental) particles that are the building blocks of matter. Elementary particles are classified as fermions and bosons, and fermions are subdivided into leptons and quarks. A lepton is a fermion that does not experience the strong interaction (or strong nuclear force), which involves coupling with the bosons known as gluons. In other words, leptons are those fermions that "ignore" gluons. By comparison, quarks are fermions that couple with gluons to form composite particles such as protons and neutrons.

Leptons and quarks come in pairs, and in three generations. Everyday matter is composed of the first generation: two leptons, the electron and electron-neutrino; and two quarks, called Up and Down.

As is the case for all fundamental particles, the lepton has properties of both a wave and a particle—it exhibits what is known as "wave-particle duality." The usual convention is to refer to such unified wave-particle fundamental entities as just "particles." The particle aspect is point-like even at scales thousands of times smaller than the proton size.

Properties of Leptons

As is the case for all fundamental particles, the lepton is a unified entity of wave and particle—the wave-particle duality of quantum physics. The wave "tells" the particle what to do over time, while the interactions of the particle "tell" the wave how to develop and resonate. The particle aspect is point-like even at scales thousands of times smaller than the proton size. The usual convention is to refer to such unified wave-particle fundamental entities as just 'particles'.

There are three known flavors of lepton: the electron, the muon, and the tau. Each flavor is represented by a pair of particles called a weak doublet. One is a massive charged particle that bears the same name as its flavor (like the electron). The other is a nearly massless neutral particle called a neutrino (such as the electron neutrino). All six of these particles have corresponding antiparticles (such as the positron or the electron antineutrino). All known charged leptons have a single unit of negative or positive electric charge (depending on whether they are particles or antiparticles) and all of the neutrinos and antineutrinos have zero electric charge. The charged leptons have two possible spin states, while only one helicity is observed for the neutrinos (all the neutrinos are left-handed, and all the antineutrinos are right-handed).

The masses of the leptons also obey a simple relation, known as the Koide formula, but at present this relationship cannot be explained.

When particles interact, generally the number of leptons of the same type (electrons and electron neutrinos, muons and muon neutrinos, tau leptons and tau neutrinos) remains the same. This principle is known as conservation of lepton number. Conservation of the number of leptons of different flavors (for example, electron number or muon number) may sometimes be violated (as in neutrino oscillation). A much stronger conservation law is the total number of leptons of all flavors, which is violated by a tiny amount in the Standard Model by the so-called chiral anomaly.

The couplings of the leptons to gauge bosons are flavor-independent. This property is called lepton universality and has been tested in measurements of the tau and muon lifetimes and of Z-boson partial decay widths, particularly at the SLC and LEP experiments.

Quantum Spin

Fermions and bosons are distinguished by their quantum spin and the type of quantum probability statistics they obey: Fermi-Dirac probability or Bose-Einstein probability, neither of which is like classical probability. (This is a rough illustration of the difference: (one) The probability of two classical coins coming up the same side—HH or TT—is 50 percent. (two) For two boson coins, the probability of such a pair is 100 percent. (three) For two fermion coins, the probability of a pair is exactly zero percent, it is forbidden, and you always get HT. Fermions are said to have quantum spin -½, giving them the odd property of having to be rotated 720° in order to get back to where you started. (A familiar example of this sort of behavior is the Moebius Strip.) Bosons have quantum spin -1, and take the usual 360° to rotate back to where they started.

Table of the Leptons:

Charged lepton / antiparticle				Neutrino / antineutrino			
Name	Symbol	Electric charge (e)	Mass (MeV/ c^2)	Name	Symbol	Electric charge (e)	Mass (MeV/c^2)
Electron/ Positron	e^- / e^+	−1 / +1	0.511	Electron neutrino / Electron antineutrino	$\nu_e / \overline{\nu}_e$	0	< 0.0000022
Muon	μ^- / μ^+	−1 / +1	105.7	Muon neutrino / Muon antineutrino	$\nu_\mu / \overline{\nu}_\mu$	0	< 0.17
Tau lepton	τ^- / τ^+	−1 / +1	1777	Tau neutrino / Tau antineutrino	$\nu_\tau / \overline{\nu}_\tau$	0	< 15.5

Note that the neutrino masses are known to be non-zero because of neutrino oscillation, but their masses are sufficiently light that they have not been measured directly as of 2007. The names "mu" and "tau" seem to have been selected due to their places in the Greek alphabet; mu is seven letters after epsilon (electron), whereas tau is seven letter after mu.

Muon

The muon is an elementary particle similar to the electron, with an electric charge of −1 *e* and a spin of 1/2, but with a much greater mass. It is classified as a lepton. As is the case with other leptons, the muon is not believed to have any sub-structure—that is, it is not thought to be composed of any simpler particles.

The muon is an unstable subatomic particle with a mean lifetime of 2.2 μs, much longer than many other subatomic particles. As with the decay of the non-elementary neutron (with a lifetime around 15 minutes), muon decay is slow (by subatomic standards) because the decay is mediated by the weak interaction exclusively (rather than the more powerful strong interaction or electromagnetic interaction), and because the mass difference between the muon and the set of its decay products is small, providing few kinetic degrees of freedom for decay. Muon decay almost always produces at least three particles, which must include an electron of the same charge as the muon and two neutrinos of different types.

Like all elementary particles, the muon has a corresponding antiparticle of opposite charge (+1 *e*) but equal mass and spin: the antimuon (also called a *positive muon*). Muons are denoted by μ^- and antimuons by μ^+. Muons were previously called mu mesons, but are not classified as mesons by modern particle physicists, and that name is no longer used by the physics community.

Muons have a mass of 105.66 MeV/c^2, which is about 207 times that of the electron. Due to their greater mass, muons are not as sharply accelerated when they encounter electromagnetic fields, and do not emit as much bremsstrahlung (deceleration radiation). This allows muons of a given energy to penetrate far more deeply into matter than electrons since the deceleration of electrons and muons is primarily due to energy loss by the bremsstrahlung mechanism. As an example, so-called "secondary muons", generated by cosmic rays hitting the atmosphere, can penetrate to the Earth's surface, and even into deep mines.

Because muons have a very large mass and energy compared with the decay energy of radioactivity, they are never produced by radioactive decay. They are, however, produced in copious amounts in high-energy interactions in normal matter, in certain particle accelerator experiments with hadrons, or naturally in cosmic ray interactions with matter. These interactions usually produce pi mesons initially, which most often decay to muons.

As with the case of the other charged leptons, the muon has an associated muon neutrino, denoted by ν_μ, which is not the same particle as the electron neutrino, and does not participate in the same nuclear reactions.

Muon Sources

Muons arriving on the Earth's surface are created indirectly as decay products of collisions of cosmic rays with particles of the Earth's atmosphere.

About 10,000 muons reach every square meter of the earth's surface a minute; these charged particles form as by-products of cosmic rays colliding with molecules in the upper atmosphere. Traveling at relativistic speeds, muons can penetrate tens of meters into rocks and other matter before attenuating as a result of absorption or deflection by other atoms.

When a cosmic ray proton impacts atomic nuclei in the upper atmosphere, pions are created. These decay within a relatively short distance (meters) into muons (their preferred decay product), and muon neutrinos. The muons from these high energy cosmic rays generally continue in about the same direction as the original proton, at a velocity near the speed of light. Although their lifetime *without* relativistic effects would allow a half-survival distance of only about 456 m (2.197 μs×ln(2) × 0.9997×c) at most (as seen from Earth) the time dilation effect of special relativity (from the viewpoint of the Earth) allows cosmic ray secondary muons to survive the flight to the Earth's surface, since in the Earth frame the muons have a longer half life due to their velocity. From the viewpoint (inertial frame) of the muon, on the other hand, it is the length contraction effect of special relativity which allows this penetration, since in the muon frame its lifetime is unaffected, but the length contraction causes distances through the atmosphere and Earth to be far shorter than these distances in the Earth rest-frame. Both effects are equally valid ways of explaining the fast muon's unusual survival over distances.

Since muons are unusually penetrative of ordinary matter, like neutrinos, they are also detectable deep underground (700 meters at the Soudan 2 detector) and underwater, where they form a major part of the natural background ionizing radiation. Like cosmic rays, as noted, this secondary muon radiation is also directional.

The same nuclear reaction (i.e. hadron-hadron impacts to produce pion beams, which then quickly decay to muon beams over short distances) is used by particle physicists to produce muon beams, such as the beam used for the muon $g-2$ experiment.

Muon Decay

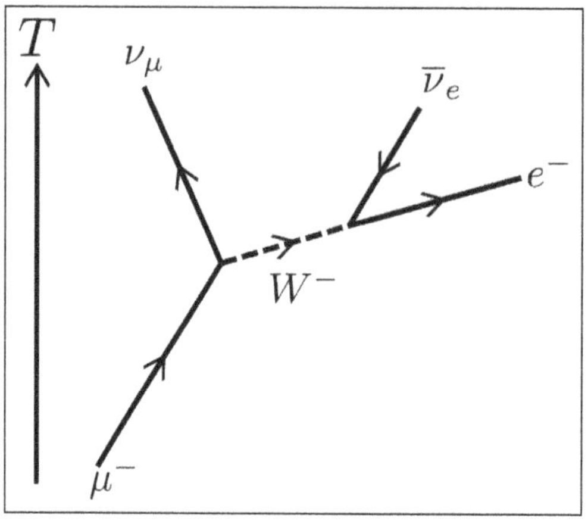

The most common decay of the muon.

Muons are unstable elementary particles and are heavier than electrons and neutrinos but lighter than all other matter particles. They decay via the weak interaction. Because leptonic family numbers are conserved in the absence of an extremely unlikely immediate neutrino oscillation, one of the product neutrinos of muon decay must be a muon-type neutrino and the other an electron-type antineutrino. Because charge must be conserved, one of the products of muon decay is always an electron of the same charge as the muon (a positron if it is a positive muon). Thus all muons decay to at least an electron, and two neutrinos. Sometimes, besides these necessary products, additional other particles that have no net charge and spin of zero (e.g., a pair of photons, or an electron-positron pair), are produced.

The dominant muon decay mode (sometimes called the Michel decay after Louis Michel) is the simplest possible: the muon decays to an electron, an electron antineutrino, and a muon neutrino. Antimuons, in mirror fashion, most often decay to the corresponding antiparticles: a positron, an electron neutrino, and a muon antineutrino. In formulaic terms, these two decays are:

$$\mu^- \rightarrow e^- + \overline{\nu}_e + \nu_\mu$$

$$\mu^+ \rightarrow e^+ + \overline{\nu}_e + \nu_\mu$$

The mean lifetime, $\tau = \hbar/\Gamma$, of the (positive) muon is (2.196 9811 ± 0.000 0022) μs. The equality of the muon and antimuon lifetimes has been established to better than one part in 10^4.

Prohibited Decays

Certain neutrino-less decay modes are kinematically allowed but are, for all practical purposes, forbidden in the Standard Model, even given that neutrinos have mass and oscillate. Examples forbidden by lepton flavour conservation are:

$$\mu^- \rightarrow e^- + \gamma$$

and

$$\mu^- \rightarrow e^- + e^+ + e^-.$$

To be precise in the Standard Model with neutrino mass, a decay like:

$$\mu^- \rightarrow e^- + \gamma$$

is technically possible, for example by neutrino oscillation of a virtual muon neutrino into an electron neutrino, but such a decay is astronomically unlikely and therefore should be experimentally unobservable: less than one in 10^{50} muon decays should produce such a decay.

Observation of such decay modes would constitute clear evidence for theories beyond the Standard Model. Upper limits for the branching fractions of such decay modes were measured in many experiments starting more than 50 years ago. The current upper limit for the $\mu^+ \rightarrow e^+ + \gamma$ branching fraction was measured 2009–2013 in the MEG experiment and is 4.2×10^{-13}.

Theoretical Decay Rate

The muon decay width which follows from Fermi's golden rule has dimension of energy, and must be proportional to the square of the amplitude, and thus the square of Fermi's coupling constant (G_F), with over-all dimension of inverse fourth power of energy. By dimensional analysis, this leads to Sargent's rule of fifth-power dependence on m_μ,

$$\Gamma = \frac{G_F^2 m_\mu^5}{192\pi^3} I\left(\frac{m_e^2}{m_\mu^2}\right),$$

where $I(x) = 1 - 8x - 12x^2 \ln x + 8x^3 - x^4$,, and $x = 2E_e / m_\mu c^2$ is the fraction of the maximum energy transmitted to the electron.

The decay distributions of the electron in muon decays have been parameterised using the so-called Michel parameters. The values of these four parameters are predicted unambiguously in the Standard Model of particle physics, thus muon decays represent a good test of the space-time structure of the weak interaction. No deviation from the Standard Model predictions has yet been found.

For the decay of the muon, the expected decay distribution for the Standard Model values of Michel parameters is:

$$\frac{d^2\Gamma}{dx\,d\cos\theta} \sim x^2[(3-2x) + P_\mu \cos\theta(1-2x)]$$

where θ is the angle between the muon's polarization vector P_μ and the decay-electron momentum vector, and $\mathrm{P}_\mu = |\mathrm{P}_\mu|$ is the fraction of muons that are forward-polarized. Integrating this expression over electron energy gives the angular distribution of the daughter electrons:

$$\frac{d\Gamma}{d\cos\theta} \sim 1 - \frac{1}{3} P_\mu \cos\theta.$$

The electron energy distribution integrated over the polar angle (valid for $x < 1$) is,

$$\frac{d\Gamma}{dx} \sim (3x^2 - 2x^3).$$

Due to the muons decaying by the weak interaction, parity conservation is violated. Replacing the $\cos\theta$ term in the expected decay values of the Michel Parameters with a $\cos\omega t$ term, where ω is the Larmor frequency from Larmor precession of the muon in a uniform magnetic field, given by:

$$\omega = \frac{egB}{2m}$$

where m is mass of the muon, e is charge, g is the muon g-factor and B is applied field.

A change in the electron distribution computed using the standard, unprecessional, Michel Parameters can be seen displaying a periodicity of π radians. This can be shown to physically correspond to a phase change of π, introduced in the electron distribution as the angular momentum is changed by the action of the charge conjugation operator, which is conserved by the weak interaction.

The observation of parity violation in muon decay can be compared to the concept of violation of parity in weak interactions in general as an extension of The Wu Experiment, as well as the change of angular momentum introduced by a phase change of π corresponding to the charge-parity operator being invariant in this interaction. This fact is true for all lepton interactions in The Standard Model.

Muonic Atoms

The muon was the first elementary particle discovered that does not appear in ordinary atoms.

Negative Muon Atoms

Negative muons can, however, form muonic atoms (previously called mu-mesic atoms), by replacing an electron in ordinary atoms. Muonic hydrogen atoms are much smaller than typical hydrogen atoms because the much larger mass of the muon gives it a much more localized ground-state wavefunction than is observed for the electron. In multi-electron atoms, when only one of the electrons is replaced by a muon, the size of the atom continues to be determined by the other electrons, and the atomic size is nearly unchanged. However, in such cases the orbital of the muon continues to be smaller and far closer to the nucleus than the atomic orbitals of the electrons.

Muonic helium is created by substituting a muon for one of the electrons in helium-4. The muon orbits much closer to the nucleus, so muonic helium can therefore be regarded like an isotope of helium whose nucleus consists of two neutrons, two protons and a muon, with a single electron outside. Colloquially, it could be called "helium 4.1", since the mass of the muon is slightly greater than 0.1 amu. Chemically, muonic helium, possessing an unpaired valence electron, can bond with other atoms, and behaves more like a hydrogen atom than an inert helium atom.

Muonic heavy hydrogen atoms with a negative muon may undergo nuclear fusion in the process of muon-catalyzed fusion, after the muon may leave the new atom to induce fusion in another hydrogen molecule. This process continues until the negative muon is captured by a helium nucleus, and cannot escape until it decays.

Finally, a possible fate of negative muons bound to conventional atoms is that they are captured by the weak-force by protons in nuclei in a sort of electron-capture-like process. When this happens, the proton becomes a neutron and a muon neutrino is emitted.

Positive Muon Atoms

A *positive* muon, when stopped in ordinary matter, cannot be captured by a proton since it would need to be an antiproton. The positive muon is also not attracted to the nucleus of atoms. Instead, it binds a random electron and with this electron forms an exotic atom known as muonium (Mu) atom. In this atom, the muon acts as the nucleus. The positive muon, in this context, can be considered a pseudo-isotope of hydrogen with one ninth of the mass of the proton. Because the reduced mass of muonium, and hence its Bohr radius, is very close to that of hydrogen, this short-lived "atom" (or a muon and electron) behaves chemically—to a first approximation—like the isotopes of hydrogen (protium, deuterium and tritium).

Both positive and negative muons can be part of a short-lived pi-mu atom consisting of a muon and an oppositely charged pion. These atoms were observed in the 1970s in experiments at Brookhaven and Fermilab.

Use in Measurement of the Proton Charge Radius

The experimental technique that is expected to provide the most precise determination of the root-mean-square charge radius of the proton is the measurement of the frequency of photons (precise "color" of light) emitted or absorbed by atomic transitions in muonic hydrogen. This form of hydrogen atom is composed of a negatively charged muon bound to a proton. The muon is particularly well suited for this purpose because its much larger mass results in a much more compact bound state and hence a larger probability for it to be found inside the proton in muonic hydrogen compared to the electron in atomic hydrogen. The Lamb shift in muonic hydrogen was measured by driving the muon from a 2s state up to an excited 2p state using a laser. The frequency of the photons required to induce two such (slightly different) transitions were reported in 2014 to be 50 and 55 THz which, according to present theories of quantum electrodynamics, yield an appropriately averaged value of 0.84087 ± 0.00039 fm for the charge radius of the proton.

The internationally accepted value of the proton's charge radius is based on a suitable average of results from older measurements of effects caused by the nonzero size of the proton on scattering of electrons by nuclei and the light spectrum (photon energies) from excited atomic hydrogen. The official value updated in 2014 is 0.8751 ± 0.0061 fm. The expected precision of this result is inferior to that from muonic hydrogen by about a factor of fifteen, yet they disagree by about 5.6 times the nominal uncertainty in the difference (a discrepancy called 5.6σ in scientific notation). A conference of the world experts on this topic led to the decision to exclude the muon result from influencing the official 2014 value, in order to avoid hiding the mysterious discrepancy. This "proton radius puzzle" remained unresolved as of late 2015, and has attracted much attention, in part because of the possibility that both measurements are valid, which would imply the influence of some "new physics".

Anomalous Magnetic Dipole Moment

The anomalous magnetic dipole moment is the difference between the experimentally observed value of the magnetic dipole moment and the theoretical value predicted by the Dirac equation. The measurement and prediction of this value is very important in the precision tests of QED (quantum electrodynamics). The E821 experiment at Brookhaven National Laboratory (BNL) studied the precession of muon and anti-muon in a constant external magnetic field as they circulated in a confining storage ring. E821 reported the following average value in 2006:

$$a = \frac{g-2}{2} = 0.00116592080(54)(33)$$

where the first errors are statistical and the second systematic.

The prediction for the value of the muon anomalous magnetic moment includes three parts:

$$a_{\mu}^{SM} = a_{\mu}^{QED} + a_{\mu}^{EW} + a_{\mu}^{had}.$$

The difference between the g-factors of the muon and the electron is due to their difference in mass. Because of the muon's larger mass, contributions to the theoretical calculation of its anomalous magnetic dipole moment from Standard Model weak interactions and from contributions involving hadrons are important at the current level of precision, whereas these effects are not important for the electron. The muon's anomalous magnetic dipole moment is also sensitive to contributions from new physics beyond the Standard Model, such as supersymmetry. For this reason, the muon's anomalous magnetic moment is normally used as a probe for new physics beyond the Standard Model rather than as a test of QED. Muon $g-2$, a new experiment at Fermilab using the E821 magnet will improve the precision of this measurement.

Muon Radiography and Tomography

Since muons are much more deeply penetrating than X-rays or gamma rays, muon imaging can be used with much thicker material or, with cosmic ray sources, larger objects. One example is commercial muon tomography used to image entire cargo containers to detect shielded nuclear material, as well as explosives or other contraband.

The technique of muon transmission radiography based on cosmic ray sources was first used in the 1950s to measure the depth of the overburden of a tunnel in Australia and in the 1960s to search for possible hidden chambers in the Pyramid of Chephren in Giza. In 2017, the discovery of a large void (with a length of 30 m minimum) by observation of cosmic-ray muons was reported.

In 2003, the scientists at Los Alamos National Laboratory developed a new imaging technique: muon scattering tomography. With muon scattering tomography, both incoming and outgoing trajectories for each particle are reconstructed, such as with sealed aluminum drift tubes. Since the development of this technique, several companies have started to use it.

In August 2014, Decision Sciences International Corporation announced it had been awarded a contract by Toshiba for use of its muon tracking detectors in reclaiming the Fukushima nuclear complex. The Fukushima Daiichi Tracker (FDT) was proposed to make a few months of muon measurements to show the distribution of the reactor cores.

In December 2014, Tepco reported that they would be using two different muon imaging techniques at Fukushima, "Muon Scanning Method" on Unit 1 (the most badly damaged, where the fuel may have left the reactor vessel) and "Muon Scattering Method" on Unit 2.

The International Research Institute for Nuclear Decommissioning IRID in Japan and the High Energy Accelerator Research Organization KEK call the method they developed for Unit 1 the muon permeation method; 1,200 optical fibers for wavelength conversion light up when muons come into contact with them. After a month of data collection, it is hoped to reveal the location and amount of fuel debris still inside the reactor. The measurements began in February 2015.

Bosons

In particle physics, a boson is a type of particle that obeys the rules of Bose-Einstein statistics. These bosons also have a quantum spin with contains an integer value, such as 0, 1, -1, -2, 2, etc

Bosons are sometimes called force particles, because it is the bosons that control the interaction of physical forces, such as electromagnetism and possibly even gravity itself.

The name boson comes from the surname of Indian physicist Satyendra Nath Bose, a brilliant physicist from the early twentieth century who worked with Albert Einstein to develop a method of analysis called Bose-Einstein statistics. In an effort to fully understand Planck's law (the thermodynamics equilibrium equation that came out of Max Planck's work on the blackbody radiation problem), Bose first proposed the method in a 1924 paper trying to analyze the behavior of photons. He sent the paper to Einstein, who was able to get it published and then went on to extend Bose's reasoning beyond mere photons, but also to apply to matter particles.

One of the most dramatic effects of Bose-Einstein statistics is the prediction that bosons can overlap and coexist with other bosons. Fermions, on the other hand, cannot do this, because they follow the Pauli Exclusion Principle (chemists focus primarily on the way the Pauli Exclusion Principle impacts the behavior of electrons in orbit around an atomic nucleus.) Because of this, it is possible for photons to become a laser and some matter is able to form the exotic state of a Bose-Einstein condensate.

Fundamental Bosons

According to the Standard Model of quantum physics, there are a number of fundamental bosons, which are not made up of smaller particles. This includes the basic gauge bosons, the particles that mediate the fundamental forces of physics (except for gravity, which we'll get to in a moment). These four gauge bosons have spin 1 and have all been experimentally observed:

- Photon - Known as the particle of light, photons carry all electromagnetic energy and act as the gauge boson that mediates the force of electromagnetic interactions.

- Gluon - Gluons mediate the interactions of the strong nuclear force, which binds together quarks to form protons and neutrons and also holds the protons and neutrons together within an atom's nucleus.
- W Boson - One of the two gauge bosons involved in mediating the weak nuclear force.
- Z Boson - One of the two gauge bosons involved in mediating the weak nuclear force.

In addition to the above, there are other fundamental bosons predicted, but without clear experimental confirmation (yet):

- Higgs Boson - According to the Standard Model, the Higgs Boson is the particle that gives rise to all mass. On July 4, 2012, scientists at the Large Hadron Collider announced that they had good reason to believe they'd found evidence of the Higgs Boson. Further research is ongoing in an attempt to get better information about the particle's exact properties. The particle is predicted to have a quantum spin value of 0, which is why it is classified as a boson.
- Graviton - The graviton is a theoretical particle which has not yet been experimentally detected. Since the other fundamental forces - electromagnetism, strong nuclear force, and weak nuclear force - are all explained in terms of a gauge boson that mediates the force, it was only natural to attempt to use the same mechanism to explain gravity. The resulting theoretical particle is the graviton, which is predicted to have a quantum spin value of 2.
- Bosonic Superpartners - Under the theory of supersymmetry, every fermion would have a so-far-undetected bosonic counterpart. Since there are 12 fundamental fermions, this would suggest that - if supersymmetry is true - there are another 12 fundamental bosons that have not yet been detected, presumably because they are highly unstable and have decayed into other forms.

Composite Bosons

Some bosons are formed when two or more particles join together to create an integer-spin particle, such as:

- Mesons - Mesons are formed when two quarks bond together. Since quarks are fermions and have half-integer spins, if two of them are bonded together, then the spin of the resulting particle (which is the sum of the individual spins) would be an integer, making it a boson.
- Helium-4 atom - A helium-4 atom contains 2 protons, 2 neutrons, and 2 electrons and if you add up all of those spins, you'll end up with an integer every time. Helium-4 is particularly noteworthy because it becomes a superfluid when cooled to ultra-low temperatures, making it a brilliant example of Bose-Einstein statistics in action.

If you're following the math, any composite particle that contains an even number of fermions is going to be a boson, because an even number of half-integers is always going to add up to an integer.

The Quark Model

Isospin Symmetry

In early studies of nuclear reactions, it was found that, to a good approximation, nuclear force is independent of the electromagnetic charge carried by the nucleons ó charge independence. In other words, strong interaction has an $SU(2)$ symmetry which transforms n into p and vice versa. These $SU(2)$ generators T_1; T_2; T_3 satisfy the commutation similar to that of angular momenta,

$$\left[\mathrm{T_i}\ ;\ \mathrm{T_j}\right] = \mathrm{i} \in_{\mathrm{ijk}} \mathrm{T_k}$$

Acting on n or p we have,

$$T_3 \,|\, p\rangle = \frac{1}{2}|\, p\rangle\,,\ T_3 \,|\, n\rangle = -\frac{1}{2}\ |\, n\rangle; T+|\, n\rangle = |p\rangle\ T_{-}|p\rangle = |n\rangle$$

This means that n or p form a doublet under isospin transformation. Isospin invariance simply means that,

$$[T_i\ , H_s] = 0$$

where Hs is the strong interation Hamiltonian. We can extend the isospin assignments to other hadrons by assuming isospin invariant in their productions. For example we get the following isospin multiplets,

$$\left(\pi^+;\ \pi^0\ ;\ \pi^-\right) I = 1;\ \left(K^+;\ K^0\right);\ \left(\overline{K}^0;\ K^-\right) I = \frac{1}{2}\ ;\ \eta\ ;\ I = 0$$

$$\left(\Sigma^+;\ \Sigma^0\ ;\ \Sigma^-\right) I = 1, \left(\Xi^0;\ \Xi^-\right) I = \frac{1}{2}\ , \Delta,\ I = 0$$

$$\left(\rho^+;\ \rho^0\ ;\ \rho^-\right) I = 1, \left(K^{+*};\ \mathrm{K}^{0-}\right), \left(\overline{K}^{0*};\ \mathrm{K}^{*-}\right) I = \frac{1}{2}$$

If isospin symmetry were exact, then all particles in the same multiplets have same masses, which is not the case in nature. But the mass difference within the isospin multiplets seems to be quite small.

$$\frac{m_n - m_p}{m_n + m_p} \sim 0.7\times10^{-3}, \frac{m_{\pi+} - m_\pi 0}{m_{\pi+} + m_\pi 0} \sim 1.7\times10^{-2}$$

Thus we can treat the isospin symmetry as approximate one and maybe it is good to few%.

SU(3) symmetry and Quark Model

When and K particles were discovered, they were produced in pair (associated production) with longer life time. It was postulated that these new particles possessed a new additive quantum number, called strangeness S, conserved by strong interaction but violated in decays,

$$s(\wedge^0) = -1, s(k^0) = 1$$

Extension to other hadrons systematically, we can get a general relation,

$$Q = T_3 + \frac{Y}{2}$$

where $Y = B + S$ is called hyperchargee; and B is the baryon number. This is known as Gell-Mann-Nishijima relation.

Eight-fold way: Gell-Mann, Neeman When we group mesons or baryons with same spin and parity,

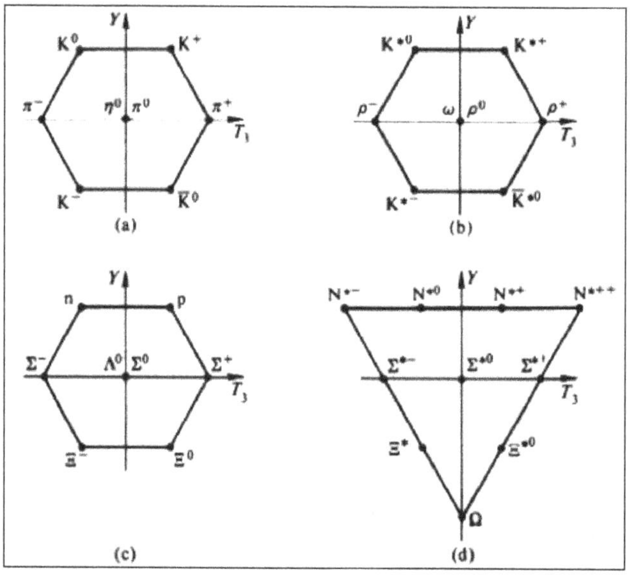

we see that they form a patterns of 8 or 10 as shown in the figure. These are the same as some irreducible representations of $SU(3)$ group. This can be extended to other hadrons which are either in octet or decouplet representations of $SU(3)$ Thus the spectra of hadrons seem to show some pattern of $SU(3)$ symmetry. But this symmetry is lots worse than isospin symmetry of $SU(2)$ because the mass splitting within the $SU(3)$) multiplets is about 20% at best. Nevertheless, it is still useful to classify hadrons in terms of $SU(3)$ symmetry. This is known as the eight-fold way.

Quark Model

One peculiar feature of the eight fold way is that octet and decuplet are not the fundamental representation of $SU(3)$ group. In 1964, Gell-mann and Zweig independently propose the quark model, in which all hadrons are built out of spin $\frac{1}{2}$ quarks which transform as members of the fundamental representation of $SU(3)$ the triplet,

$$q_i = \begin{pmatrix} q1 \\ q2 \\ q3 \end{pmatrix} = \begin{pmatrix} u \\ d \\ s \end{pmatrix}$$

with the quantum numbers,

	Q	T	T_3	Y	S	B
μ	2/3	1/2	+1/2	1/3	0	1/3
d	−1/3	1/2	−1/2	1/3	0	1/3
s	−1/3	0	0	−2/3	−1	1/3

In this scheme, mesons are $\overline{qq}$ bound states. For examples,

$$\pi^+ \sim \bar{d}u \; \pi^0 \sim \frac{1}{\sqrt{2}}(\bar{u}u - \bar{d}d).\pi^- \sim \bar{u}d$$

$$k^+ \sim \bar{s}u \; k^o \sim \bar{s}d, k{-} \sim \bar{u}s \; \eta^0 \sim \frac{1}{\sqrt{6}}(\bar{u}u + \bar{d}d - 2_{\bar{s}s})$$

and baryons are qqq bound states,

$$p \sim uud, n \sim ddu$$

$$\Sigma^+ \sim suu, \Sigma^0 \sim s\left(\frac{ud + du}{\sqrt{2}}\right), \Sigma^- \sim sdd$$

$$\Xi^0 \sim ssu, \Xi^- \sim ssd, \wedge^0 \sim \frac{s(ud - du)}{\sqrt{2}}$$

It seems that the quantum numbers of the hadrons are all carried by the quarks. But we do not know the dynamics which bound the quarks into hadrons. Since quarks are the fundamental constituent of hadrons it is important to Önd these particles. But over the years none have been found.

Paradoxes of Simple Quark Model

- Quarks have fractional charges while all observed particles have integer charges. At least one of the quarks is stable. None has been found.
- Hadrons are exclusively made out $\bar{q}q$; qqq bound states. In other word, qq; $qqqq$ states are absent.
- The quark content of the baryon N^{*++} is uuu: If we chose the spin state to be $\left|\frac{3}{2}, \frac{3}{2}\right\rangle$ then all quarks are in spin-up state~ $^{\alpha}1^{\alpha}2^{\alpha}3$ which is totally symmetric: If we assume that the ground state has l = 0, then spatical wave function is also symmetric. This will leads to violation of Pauli exclusiion principle.

Color Degree of Freedom

One way to get out of these problems, is to introduce color degrees of freedom for each quark and postulates that only color singlets are physical observables. 3 colors are needed to get antisymmetric wave function for N^{*++} and remains a color singlet state. In other words each quark comes in 3 colors,

$$u_{\alpha} = (u_1, u_2, u_3),\ d_{\alpha} = (d_1, d_2, d_3)$$

All hadrons form singlets under $SU(3)_{color}$ symmetry, e.g.

$$N^{*++} \sim u_{\alpha}(x_1) \propto_{\beta} (x_2) u_{\gamma}(x_3) \varepsilon^{\alpha\beta\gamma}$$

Futhermore, color singlets can not be formed from the combination qq; $qqqq$ and they are absent from the observed specrum. Needless to say that a single quark is not observable.

Gell-Mann Okubo Mass Formula

Since $SU(3)$ is not an exact symmetry, we want to see whether we can understand the pattern of the $SU(3)$ breaking. Experimentally $SU(2)$ seems to be a good symmetry, we will assume isospin symmetry to set $m_u = m_d$. We will assume that we can write the hadron masses as linear combinations of quark masses.

- 0^- mesons

Here we assume that the meson masses are linear functions of quark masses,

$$m_{\pi}^2 = \lambda(m_o + 2m_u)$$

$$m_k^2 = \lambda(m_o + 2m_u + m_s)$$

$$m_{\eta}^2 = \lambda\left[m_o + \frac{2}{3}(m_u + 2m_s)\right]$$

where λ and m_o are some constants with mass dimension. Eliminate the quark masses we get,

$$4m_k^2 = m_{\pi}^2 + 3m_{\eta}^2$$

This known as the Gell-Mann Okubo mass formula. Experimentally, we hav $LHS = 4m_k^2 \simeq 0{:}98(Gev)^2$ while $RHS = m_{\pi}^2 + 3m^2\ ' \simeq 0{:}92(Gev)^2$. This seems to show that this formula works quite well.

- $\frac{1}{2}^+$ *baryon*

The masses of $\frac{1}{2}^+$ baryons can be written as,

$$mN = m_o + 3m_u$$

$$m\Sigma = m_o + 2m_u + M_S$$

$$m\Xi = m_o + m_u + 2M_S$$

$$m_{\wedge} = m_o + 2m_u + M_S$$

The Gell-Mann-Okubo mass formula takes the form,

$$\frac{m_{\Sigma} + 3m_{\wedge}}{2} \simeq mN + m\Xi$$

Expermentally, $\frac{m_{\Sigma} + 3m_{\wedge}}{2} \simeq 2.23 Gev$, and $m_N + m\Xi \simeq 2.25 Gev$

- $\frac{3}{2}^{+}$ *baryon*

The mass relation here is quite simple,

$$m\Omega - m\Xi^* = m\Xi^* \; m\Sigma^* = mN^*$$

This sometimes is referred to as equal spacing rule. In fact when this relation is derived the particle.

Ω has not yet been found and this relation is used to predicted the mass of Ω and subsequent discovery gives a very strong support to the idea of $SU\ (3)$ symmetry.

ω − φ Mixing

For the 1^- mesons, the situation seems to be quite di§erent. If we make an analogy with 0^- mesons, we would get the Gell-Mann Okubo mass relation in the form,

$$3m_{\omega}^2 = 4m_{K*}^2 - m_p^2$$

Using $m_K{}^* = 890 mev$ and $m_\rho = 770 mev$ we would get $m_\omega = 926{:}5$ mev from this. But experimentally, we have $m_\omega = 783 mev$ which is qiute far away. On the other hand, there is a meson with mass $m_\varphi = 102 mev$ +and has same $SU\ (2)$ quatntum number as ω In principle, when $SU\ (3)$ symmetry is broken, $\omega - \varphi$ mixing is possible. Suppose for some reason there is a signiÖcant $\omega - \varphi$ mixing we want to see whether this can save the mass relation.

Denote the $SU\ (3)$ octet state by v_8 and singlet state by V1,

$$v_s = \frac{1}{\sqrt{6}}(u\bar{u} + d\bar{d} - 2s\bar{s})$$

$$v_1 = \frac{1}{\sqrt{3}}(u\bar{u} + d\bar{d} - s\bar{s})$$

Write the mass matrix as,

$$m = \begin{pmatrix} m_{88}^2 & m_{18}^2 \\ m_{18}^2 & m_{11}^2 \end{pmatrix}$$

Assume that the octet mass is that predicted by Gell-Mann Okubo mass.

Relation, i.e.

$$3m^2_{88} = 4m^2_{K*} - \mathrm{m}^2_{\rho}$$

After diagonalizing the mass matrix M, we get,

$$R^{+}MR = M_d = \begin{pmatrix} m^2_{\omega} & 0 \\ 0 & m^2_{\varphi} \end{pmatrix}$$

with,

$$R = \begin{pmatrix} \cos\theta & \sin\theta \\ -\sin\theta & \cos\theta \end{pmatrix}$$

thus,

$$\begin{pmatrix} \cos\theta & \sin\theta \\ -\sin\theta & \cos\theta \end{pmatrix}\begin{pmatrix} m_{\omega}^2 & 0 \\ 0 & m_{\omega}^2 \end{pmatrix}\begin{pmatrix} \cos\theta & -\sin\theta \\ \sin\theta & \cos\theta \end{pmatrix}$$

$$= \begin{pmatrix} m^2_{\varphi}\sin^2\theta + m^2_{\omega}\cos^2\theta & m^2_{\varphi}\cos\theta\sin\theta - m^2_{\omega}\cos\theta\sin\theta \\ m^2_{\varphi}\cos\theta\sin\theta - m^2_{\omega}\cos\theta\sin\theta & m^2_{\varphi}\cos^2\theta + m^2_{\omega}\sin^2\theta \end{pmatrix}$$

and we get,

$$\sin\theta = \sqrt{\left(\frac{m^2_{88} - m^2_{\omega}}{m^2_{\varphi} - m^2_{\omega}}\right)}$$

The mass eigenstates are:

$$\omega = \cos\theta V_S - \sin\theta v_1$$

$$\varphi = \sin\theta Vs + \cos\theta V_1$$

Using m_{88} = 926:5Mev from Gell-Mann Okubo mass formula, we get:

$$\sin\theta = 0.76$$

This is very close to the ideal mixing $\sin\theta = \sqrt{\frac{2}{3}}$ = 0:81 where mass eigenstates have a simple form,

$$\omega = \frac{1}{\sqrt{2}}\left(\bar{u}u + \bar{d}d\right)$$

$$\varphi = \bar{8}8$$

This means that the physical φ meson is mostly made out of s quarks in this scheme.

Zweig Rule

Since ω and have same quantum numbers under $SU(2)$ one expects they have similar decay widths. Experimentally $\omega \to 3\pi$ mostly, but $\varphi \to 3\pi$ is very suppressed relative to $\varphi \to k^+k^-$ channel even though $\varphi \to kk$ has very small phase space since m_θ = 1020 Mev and m_k ≈ 494 *Mev*,

$$B(\varphi \to KK) \approx 85\%$$

$$B(\varphi \to \pi\pi\pi) \approx 28\%$$

Quark Diagrams

In term quarks contents, the decays of φ meson proceed as following diagrams indicate ! K+K

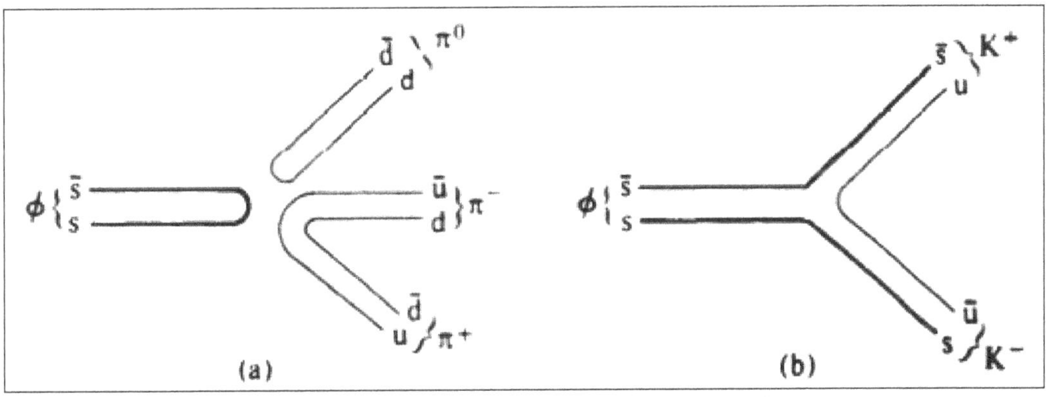

ϕ decays: (a) disalloewd; (b) allowed by Zweig rule.

Zweig rule postulates that processes involving quark-antiquark annihilation are highly suppressed for some reason. This explains why φ has a width $r_\varphi \approx$ 4:26 Mev smaller than ! $r_\omega \approx$ 8:5 Mev:

Note that the Zweig rule is very qualitative and is hard to make it more quantitive.

J/ψ and Charm Quark

In 1974 the ψ / J (3100) particle was discovered with unusually narrow width, $r \sim 70$ kev as compared $r\rho \sim 150mev$ and $r_w \sim 10MEV$: A simple explanation is that ψ / J is a bound state of $\overline{C}C$ and is below the threshold of decaying into 2 mesons containing charm quark. Thus it can only decay by cc annihilation in the initial state. By Zweig rule, these decays are highly suppressed and have very narrow width.

Flavour Particle Physics

In particle physics, flavour or flavor refers to the *species* of an elementary particle. The Standard Model counts six flavours of quarks and six flavours of leptons. They are conventionally parameterized with *flavour quantum numbers* that are assigned to all subatomic particles. They can also be described by some of the family symmetries proposed for the quark-lepton generations.

Quantum Numbers

In classical mechanics, a force acting on a point-like particle can only alter the particle's dynamical state, i.e., its momentum, angular momentum, etc. Quantum field theory, however, allows interactions that can alter other facets of a particle's nature described by non dynamical, discrete quantum numbers. In particular, the action of the weak force is such that it allows the conversion of quantum numbers describing mass and electric charge of both quarks and leptons from one discrete type to another. This is known as a flavour change, or flavour transmutation. Due to their quantum description, flavour states may also undergo quantum superposition.

In atomic physics the principal quantum number of an electron specifies the electron shell in which it resides, which determines the energy level of the whole atom. Analogously, the five flavour quantum numbers (isospin, strangeness, charm, bottomness or topness) can characterize the quantum state of quarks, by the degree to which it exhibits six distinct flavours (u, d, s, c, b, t).

Composite particles can be created from multiple quarks, forming hadrons, such as mesons and baryons, each possessing unique aggregate characteristics, such as different masses, electric charges, and decay modes. A hadron's overall flavour quantum numbers depend on the numbers of constituent quarks of each particular flavour.

Conservation Laws

All of the various charges discussed above are conserved by the fact that the corresponding charge operators can be understood as *generators of symmetries* that commute with the Hamiltonian. Thus, the eigenvalues of the various charge operators are conserved.

Absolutely conserved flavour quantum numbers are:

- Electric charge (Q),
- Weak isospin (I_3),
- Baryon number (B),
- Lepton number (L).

In some theories, such as the grand unified theory, the individual baryon and lepton number conservation can be violated, if the difference between them ($B - L$) is conserved. All other flavour quantum numbers are violated by the electroweak interactions. Strong interactions conserve all flavours.

Flavour Symmetry

If there are two or more particles which have identical interactions, then they may be interchanged without affecting the physics. Any (complex) linear combination of these two particles give the same physics, as long as the combinations are orthogonal, or perpendicular, to each other.

In other words, the theory possesses symmetry transformations such as $M\begin{pmatrix} u \\ d \end{pmatrix}$, where u and d are the two fields and M is any 2×2 unitary matrix with a unit determinant. Such matrices form a Lie group called SU(2) This is an example of flavour symmetry.

In quantum chromodynamics, flavour is a conserved global symmetry. In the electroweak theory, on the other hand, this symmetry is broken, and flavour changing processes exist, such as quark decay or neutrino oscillations.

Flavour Quantum Numbers

Leptons

All leptons carry a lepton number $L = 1$. In addition, leptons carry weak isospin, T_3, which is $-\frac{1}{2}Y$ for the three charged leptons (i.e. electron, muon and tau) and $+\frac{1}{2}$ for the three associated neutrinos. Each doublet of a charged lepton and a neutrino consisting of opposite T_3 are said to constitute one generation of leptons. In addition, one defines a quantum number called weak hypercharge, Y_W, which is −1 for all left-handed leptons. Weak isospin and weak hypercharge are gauged in the Standard Model.

Leptons may be assigned the six flavour quantum numbers: electron number, muon number, tau number, and corresponding numbers for the neutrinos. These are conserved in strong and electromagnetic interactions, but violated by weak interactions. Therefore, such flavour quantum numbers are not of great use. A separate quantum number for each generation is more useful: electronic lepton number (+1 for electrons and electron neutrinos), muonic lepton number (+1 for muons and muon neutrinos), and tauonic lepton number (+1 for tau leptons and tau neutrinos). However, even these numbers are not absolutely conserved, as neutrinos of different generations can mix; that is, a neutrino of one flavour can transform into another flavour. The strength of such mixings is specified by a matrix called the Pontecorvo–Maki–Nakagawa–Sakata matrix (PMNS matrix).

Quarks

All quarks carry a baryon number $B = \frac{1}{3}$ They also all carry weak isospin, $T_3 = \pm\frac{1}{2}$ The positive-T_3 quarks (up, charm, and top quarks) are called *up-type quarks* and negative-T_3 quarks (down, strange, and bottom quarks) are called *down-type quarks*. Each doublet of up and down type quarks constitutes one generation of quarks.

For all the quark flavour quantum numbers listed below, the convention is that the flavour charge and the electric charge of a quark have the same sign. Thus any flavour carried by a charged meson has the same sign as its charge. Quarks have the following flavour quantum numbers:

- The third component of isospin (sometimes simply *isospin*) (I_3), which has value $I_3 = \frac{1}{2}$ for the up quark and $I_3 = \frac{1}{2}$ for the down quark.

- Strangeness (S): Defined as $S = -(n_s - n_{\bar{s}})$, where n_s represents the number of strange quarks (s) and $n_{\bar{s}}$ represents the number of strange antiquarks (s). This quantum number was introduced by Murray Gell-Mann. This definition gives the strange quark a strangeness of −1 for the above-mentioned reason.
- Charm (C): Defined as $C = (n_c - n_{\bar{c}})$, where n_c represents the number of charm quarks (c) and $n_{\bar{c}}$ represents the number of charm antiquarks. The charm quark's value is +1.
- Bottomness (or *beauty*) (B'): Defined as $B' = -(n_b - n_{\bar{b}})$, where n_b represents the number of bottom quarks (b) and $n_{\bar{b}}$ represents the number of bottom antiquarks.
- Topness (or *truth*) (T): Defined as $T = (n_t - n_{\bar{t}})$, where n_t represents the number of top quarks (t) and $n_{\bar{t}}$ represents the number of top antiquarks. However, because of the extremely short half-life of the top quark (predicted lifetime of only 5×10^{-25} s), by the time it can interact strongly it has already decayed to another flavour of quark (usually to a bottom quark). For that reason the top quark doesn't hadronize, that is it never forms any meson or baryon.

These five quantum numbers, together with baryon number (which is not a flavour quantum number), completely specify numbers of all 6 quark flavours separately (as $n_q - n_{\bar{q}}$, i.e. an antiquark is counted with the minus sign). They are conserved by both the electromagnetic and strong interactions (but not the weak interaction). From them can be built the derived quantum numbers:

- Hypercharge (Y): $Y = B + S + C + B' + T$
- Electric charge: $Q = I_3 + \frac{1}{2}Y$

The terms "strange" and "strangeness" predate the discovery of the quark, but continued to be used after its discovery for the sake of continuity (i.e. the strangeness of each type of hadron remained the same); strangeness of anti-particles being referred to as +1, and particles as −1 as per the original definition. Strangeness was introduced to explain the rate of decay of newly discovered particles, such as the kaon, and was used in the Eightfold Way classification of hadrons and in subsequent quark models. These quantum numbers are preserved under strong and electromagnetic interactions, but not under weak interactions.

For first-order weak decays, that is processes involving only one quark decay, these quantum numbers (e.g. charm) can only vary by 1, that is, for a decay involving a charmed quark or antiquark either as the incident particle or as a decay byproduct, $\Delta C = \pm1$; likewise, for a decay involving a bottom quark or antiquark $\Delta B' = \pm1$. Since first-order processes are more common than second-order processes (involving two quark decays), this can be used as an approximate "selection rule" for weak decays.

A special mixture of quark flavours is an eigenstate of the weak interaction part of the Hamiltonian, so will interact in a particularly simple way with the W bosons (charged weak interactions violate flavor). On the other hand, a fermion of a fixed mass (an eigenstate of the kinetic and strong interaction parts of the Hamiltonian) is an eigenstate of flavour. The transformation from the former basis to the flavor-eigenstate/mass-eigenstate basis for quarks underlies the

Cabibbo–Kobayashi–Maskawa matrix (CKM matrix). This matrix is analogous to the PMNS matrix for neutrinos, and quantifies flavour changes under charged weak interactions of quarks.

The CKM matrix allows for CP violation if there are at least three generations.

Antiparticles and Hadrons

Flavour quantum numbers are additive. Hence antiparticles have flavour equal in magnitude to the particle but opposite in sign. Hadrons inherit their flavour quantum number from their valence quarks: this is the basis of the classification in the quark model. The relations between the hypercharge, electric charge and other flavour quantum numbers hold for hadrons as well as quarks.

Quantum Chromodynamics

Quantum chromodynamics (QCD) contains six flavours of quarks. However, their masses differ and as a result they are not strictly interchangeable with each other. The up and down flavours are close to having equal masses, and the theory of these two quarks possesses an approximate SU(2) symmetry (isospin symmetry).

Chiral Symmetry Description

Under some circumstances (for instance when the quark masses are much smaller than the chiral symmetry breaking scale of 250MeV), the masses of quarks do not meaningfully contribute to the system's behavior, and can be ignored to zeroth approximation. The simplified behavior of flavour transformations can then be successfully modeled as acting independently on the left- and right-handed parts of each quark field. This approximate description of the flavour symmetry is described by a chiral group $SU_L(N_f) \times SU_R(N_f)$.

Vector Symmetry Description

If all quarks had non-zero but equal masses, then this chiral symmetry is broken to the *vector symmetry* of the "diagonal flavour group" $SU(N_f)$, which applies the same transformation to both helicities of the quarks. This reduction of symmetry is a form of *explicit symmetry breaking*. The strength of explicit symmetry breaking is controlled by the current quark masses in QCD.

Even if quarks are massless, chiral flavour symmetry can be spontaneously broken if the vacuum of the theory contains a chiral condensate (as it does in low-energy QCD). This gives rise to an effective mass for the quarks, often identified with the valence quark mass in QCD.

Symmetries of QCD

Analysis of experiments indicate that the current quark masses of the lighter flavours of quarks are much smaller than the QCD scale, Λ_{QCD}, hence chiral flavour symmetry is a good approximation to QCD for the up, down and strange quarks. The success of chiral perturbation theory and the even more naive chiral models spring from this fact. The valence quark masses extracted from the quark model are much larger than the current quark mass. This indicates that QCD has spontaneous chiral symmetry breaking with the formation of a chiral condensate. Other phases of QCD may break the chiral flavour symmetries in other ways.

Superpartner

In particle physics, a superpartner (also sparticle) is a class of hypothetical elementary particles. Supersymmetry is one of the synergistic theories in current high-energy physics that predicts the existence of these "shadow" particles.

When considering extensions of the Standard Model, the *s-* prefix from *sparticle* is used to form names of superpartners of the Standard Model fermions (sfermions), e.g. the stop squark. The superpartners of Standard Model bosons have an *-ino* (bosinos) appended to their name, e.g. gluino, the set of all gauge superpartners are called the gauginos.

Theoretical Predictions

According to the supersymmetry theory, each fermion should have a partner boson, the fermion's superpartner, and each boson should have a partner fermion. Exact *unbroken* supersymmetry would predict that a particle and its superpartners would have the same mass. No superpartners of the Standard Model particles have yet been found. This may indicate that supersymmetry is incorrect, or it may also be the result of the fact that supersymmetry is not an exact, *unbroken* symmetry of nature. If superpartners are found, their masses would indicate the scale at which supersymmetry is broken.

For particles that are real scalars (such as an axion), there is a fermion superpartner as well as a second, real scalar field. For axions, these particles are often referred to as axinos and saxions.

In extended supersymmetry there may be more than one superparticle for a given particle. For instance, with two copies of supersymmetry in four dimensions, a photon would have two fermion superpartners and a scalar superpartner.

In zero dimensions it is possible to have supersymmetry, but no superpartners. However, this is the only situation where supersymmetry does not imply the existence of superpartners.

Recreating Superpartners

If the supersymmetry theory is correct, it should be possible to recreate these particles in high-energy particle accelerators. Doing so will not be an easy task; these particles may have masses up to a thousand times greater than their corresponding "real" particles.

Some researchers have hoped the Large Hadron Collider at CERN might produce evidence for the existence of superpartner particles. However, as of 2018, no such evidence has been found.

References

- Subatomic-particle, science: britannica.com, Retrieved 09 July, 2019
- Fermion-definition-in-physics-2699188: thoughtco.com, Retrieved 27 August, 2019
- Quark: newworldencyclopedia.org, Retrieved 13 April, 2019
- Lepton: newworldencyclopedia.org, Retrieved 05 February, 2019
- Boson-2699112: thoughtco.com, Retrieved 07 May, 2019

Quantum Electrodynamics 3

- **The Dirac Equation**
- **Solutions to the Dirac Equation**
- **The Feynman Diagrams**
- **Casimir's Trick**

Quantum electrodynamics is the quantum field theory of electrdynamics. It describes the interaction between light and matter. It includes Dirac equation, the Feynman diagrams and Casimir's trick. This chapter has been carefully written to provide an easy understanding of these aspects of quantum electrodynamics.

Quantum electrodynamics (QED) is the quantum field theory of the interactions of charged particles with the electromagnetic field. It describes mathematically not only all interactions of light with matter but also those of charged particles with one another. QED is a relativistic theory in that Albert Einstein's theory of special relativity is built into each of its equations. Because the behaviour of atoms and molecules is primarily electromagnetic in nature, all of atomic physics can be considered a test laboratory for the theory. Some of the most precise tests of QED have been experiments dealing with the properties of subatomic particles known as muons. The magnetic moment of this type of particle has been shown to agree with the theory to nine significant digits. Agreement of such high accuracy makes QED one of the most successful physical theories so far devised.

In 1928 the English physicist P.A.M. Dirac laid the foundations for QED with his discovery of a wave equation that described the motion and spin of electrons and incorporated both quantum mechanics and the theory of special relativity. The QED theory was refined and fully developed in the late 1940s by Richard P. Feynman, Julian S. Schwinger, and Tomonaga Shin'ichirō, independently of one another. QED rests on the idea that charged particles (e.g., electrons and positrons) interact by emitting and absorbing photons, the particles that transmit electromagnetic forces. These photons are "virtual"; that is, they cannot be seen or detected in any way because their existence violates the conservation of energy and momentum. The photon exchange is merely the "force" of the interaction, because interacting particles change their speed and direction of travel as they release or absorb the energy of a photon. Photons also can be emitted in a free state, in which case they may be observed as light or other forms of electromagnetic radiation.

The interaction of two charged particles occurs in a series of processes of increasing complexity. In the simplest, only one virtual photon is involved; in a second-order process, there are two; and so forth. The processes correspond to all the possible ways in which the particles can interact by the exchange of virtual photons, and each of them can be represented graphically by means of the so-called Feynman diagrams. Besides furnishing an intuitive picture of the process being considered, this type of diagram prescribes precisely how to calculate the variable involved. Each subatomic process becomes computationally more difficult than the previous one, and there are an infinite number of processes. The QED theory, however, states that the more complex the process—that is, the greater the number of virtual photons exchanged in the process—the smaller the probability of its occurrence. For each level of complexity, the contribution of the process decreases by an amount given by α^2—where α is a dimensionless quantity called the fine-structure constant, with a numerical value equal to $({}^{1}/_{137})$ Thus, after a few levels the contribution is negligible. In a more-fundamental way the factor α serves as a measure of the strength of the electromagnetic interaction. It equals $[{}^{e^2}/_{4\pi\varepsilon_o[planck]c},]$ where e is the electron charge, [planck] is Planck's constant divided by 2π, c is the speed of light, and εo is the permittivity of free space. QED is often called a perturbation theory because of the smallness of the fine-structure constant and the resultant decreasing size of higher-order contributions. This relative simplicity and the success of QED have made it a model for other quantum field theories. Finally, the picture of electromagnetic interactions as the exchange of virtual particles has been carried over to the theories of the other fundamental interactions of matter, the strong force, the weak force, and the gravitational force.

Maxwell's Equations

The Lagrangian for Maxwell's equations in the absence of any sources is simply,

$$L = -\frac{1}{4}F_{\mu\nu}F^{\mu\nu}$$

where the field strength is defined by

$$F_{\mu\nu} = \partial_\mu A_\nu - \partial_\nu A_\mu$$

The equations of motion which follow from this Lagrangian are

$$\partial\mu\left(\frac{\partial\mathcal{L}}{\partial(\partial_\mu A_\nu)}\right) = -\partial_\mu F^{\mu\nu} = 0$$

Meanwhile, from the definition o $F_{\mu\nu}$, the field strength also satisfies the Bianchi identity

$$\partial_\lambda F_{\mu\nu} + \partial_\mu F_{\nu\lambda} + \partial_\nu F_{\lambda\mu} = 0$$

To make contact with the form of Maxwell's equations you learn about in high school, we need some 3-vector notation. If we define $A_\mu = \varphi, \vec{A})$ then the electric field $\vec{E}$ and magnetic field $\vec{B}$ are defined by,

$$\vec{E} = -\nabla\varphi - \frac{\partial\vec{A}}{\partial t} \text{ and } \vec{B} = \nabla \times \vec{A}$$

which, in terms of $F_{\mu\nu}$, becomes

$$F_{\mu\nu} = \begin{pmatrix} 0 & E_x & E_y & E_z \\ -E_x & 0 & -B_z & B_y \\ -E_y & b_z & 0 & -B_x \\ -E_z & -B_y & B_x & 0 \end{pmatrix}$$

The Bianchi identity $\partial_\lambda F_{\mu\nu} + \partial_\mu F_{\nu\lambda} + \partial_\nu F_{\lambda\mu} = 0$ then gives two of Maxwell's equations,

$$\nabla \cdot \vec{B} = 0 \; and \; \frac{\partial \vec{B}}{\partial t} = -\nabla \times \vec{E}$$

These remain true even in the presence of electric sources. Meanwhile, the equations of motion give the remaining two Maxwell equations,

$$\nabla \cdot \vec{E} = 0 \; and \; \frac{\partial \vec{E}}{\partial t} = \nabla \times \vec{B}$$

Gauge Symmetry

The massless vector field A_μ has 4 components, which would naively seem to tell us that the gauge field has 4 degrees of freedom. Yet we know that the photon has only two degrees of freedom which we call its polarization states. How are we going to resolve this discrepancy? There are two related comments which will ensure that quantizing the gauge field A_μ gives rise to 2 degrees of freedom, rather than 4.

- The field A_0 has no kinetic term $\dot{A}_0$ in the Lagrangian: it is not dynamical. This means that if we are given some initial data A_i and $\dot{A}_i$ at a time t_0, then the field A_0 is fully determined by the equation of motion $\nabla \cdot \vec{E} = 0$ which, expanding out, reads,

$$\nabla^2 A_0 + \nabla \cdot \frac{\partial \vec{A}}{\partial t} = 0$$

This has the solution,

$$Ao(\vec{x}) = \int d^3x' \frac{(\nabla \cdot \partial \vec{A} / \partial t)(\vec{x}')}{4_\pi |\vec{x} - \vec{x}'|}$$

So A_0 is not independent: we don't get to specify A_0 on the initial time slice. It looks like we have only 3 degrees of freedom in A_μ rather than 4. But this is still one too many.

- The Lagrangian $\partial \mu \left(\frac{\partial \mathcal{L}}{\partial(\partial_\mu A_\nu)} \right) = -\partial_\mu F^{\mu\nu} = 0$ has a very large symmetry group, acting on the vector potential as,

$$A_\mu(x) \to A_\mu(x) + \partial_\mu \lambda(x)$$

for any function $\lambda(x)$ We'll ask only that $\lambda(x)$ dies off suitably quickly at spatial $\vec{x} \to \infty$ We call this a gauge symmetry. The field strength is invariant under the gauge symmetry:

$$F_{\mu\nu} \to \partial_\mu (A_\nu + \partial_\nu \lambda) - \partial_\nu (A_\mu + \partial_\mu \lambda) = F_{\mu\nu}$$

So what are we to make of this? We have a theory with an infinite number of symmetries, one for each function $\lambda(x)$ Previously we only encountered symmetries which act the same at all points in spacetime, for example $e\ \psi \to e^{i\alpha}\psi$ for a complex scalar field. Noether's theorem told us that these symmetries give rise to conservation laws. Do we now have an infinite number of conservation laws?

The answer is no Gauge symmetries have a very different interpretation than the global symmetries that we make use of in Noether's theorem. While the latter take a physical state to another physical state with the same properties, the gauge symmetry is to be viewed as a redundancy in our description. That is, two states related by a gauge symmetry are to be identified: they are the same physical state. One way to see that this interpretation is necessary is to notice that Maxwell's equations are not sufficient to specify the evolution of Aμ. The equations read,

$$\left[\eta_{\mu\nu} (\partial^\rho \partial_\rho) - \partial_\mu \partial_\nu \right] A^\nu = 0$$

But the operator $[\eta_{\mu\nu} (\partial^\rho \partial_\rho) - \partial_\mu \partial_\nu]$ is not invertible: it annihilates any function of the form $\partial_\mu \lambda$. This means that given any initial data, we have no way to uniquely determine A_μ at a later time since we can't distinguish between A_μ and $A_\mu + \partial_\mu \lambda$. This would be problematic if we thought that A_μ is a physical object. However, if we're happy to identify A_μ and $A_\mu + \partial_\mu \lambda$ as corresponding to the same physical state, then our problems disappear.

Since gauge invariance is a redundancy of the system, we might try to formulate the theory purely in terms of the local, physical, gauge invariant objects $\vec{E}$ and $\vec{B}$. This is fine for the free classical theory: Maxwell's equations were, after all, first written in terms of $\vec{E}$ and $\vec{B}$. But it is not possible to describe certain quantum phenomena, such as the Aharonov-Bohm effect, without using the gauge potential Aμ. We will see shortly that we also require the gauge potential to describe classically charged fields. To describe Nature, it appears that we have to introduce quantities Aμ that we can never measure.

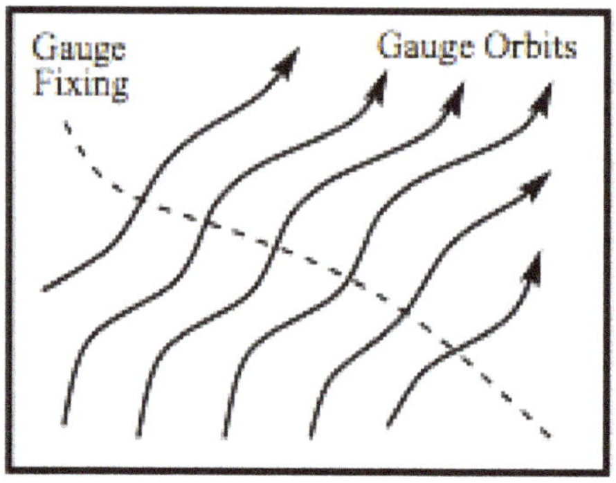

The picture that emerges for the theory of electromagnetism is of an enlarged phase space, foliated by gauge orbits as shown in the figure.

All states that lie along a given line can be reached by a gauge transformation and are identified. To make progress, we pick a representative from each gauge orbit. It doesn't matter which representative we pick — after all, they're all physically equivalent. But we should make sure that we pick a "good" gauge, in which we cut the orbits.

Different representative configurations of a physical state are called different gauges. There are many possibilities, some of which will be more useful in different situations. Picking a gauge is rather like picking coordinates that are adapted to a particular problem. Moreover,different gauges often reveal slightly different aspects of a problem. Here we'll look at two different gauges:

- Lorentz Gauge: $\partial_\mu A^\mu = 0$

To see that we can always pick a representative configuration satisfying $\partial_\mu A^\mu = 0$, suppose that we're handed a gauge field A'_μ satisfying $\partial_\mu (A')^\mu = f(x)$. Then we choose $A_\mu = A'_\mu + \partial_\mu \lambda$, where,

$$\partial_\mu \partial^\mu \lambda = -f$$

This equation always has a solution. In fact this condition doesn't pick a unique representative from the gauge orbit. We're always free to make further gauge transformations with $\partial_\mu \partial^\mu \lambda = 0$, which also has non-trivial solutions. As the name suggests, the Lorentz gauge[3] has the advantage that it is Lorentz invariant.

- Coulomb Gauge: $\nabla \cdot \vec{A} = 0$

We can make use of the residual gauge transformations in Lorentz gauge to pick $\nabla \cdot \vec{A} = 0$. (The argument is the same as before). Since A_0 is fixed by $A_0(\vec{x}) = \int d^3x' \frac{(\nabla \cdot \partial \vec{A} / \partial t)(\vec{x}')}{4_\pi |\vec{x} - \vec{x}'|}$, we have as a consequence,

$$A_0 = 0$$

(This equation will no longer hold in Coulomb gauge in the presence of charged matter). Coulomb gauge breaks Lorentz invariance, so may not be ideal for some purposes. However, it is very useful to exhibit the physical degrees of freedom: the 3 components of $\vec{A}$ satisfy a single constraint: $\nabla \cdot \vec{A} = 0$ leaving behind just 2 degrees of freedom. These will be identified with the two polarization states of the photon. Coulomb gauge is sometimes called radiation gauge.

Quantization of the Electromagnetic Field

In the following we shall quantize free Maxwell theory twice: once in Coulomb gauge, and again in Lorentz gauge. We'll ultimately get the same answers and, along the way, see that each method comes with its own subtleties.

The first of these subtleties is common to both methods and comes when computing the momentum π^μ conjugate to A_μ,

$$\pi^0 = \frac{\partial L}{\partial \dot{A}_0} = 0$$

$$\pi^i = \frac{\partial L}{\partial \dot{A}_i} = -F^{oi} \equiv E^i$$

so the momentum π^0 conjugate to A_0 vanishes. This is the mathematical consequence of the statement we made above: A_0 is not a dynamical field. Meanwhile, the momentum conjugate to A_i is our old friend, the electric field. We can compute the Hamiltonian,

$$H = \int d^3 x \pi^i \dot{A}_i - L$$

$$= \int d^3x \frac{1}{2}\vec{E}\cdot\vec{E} + \frac{1}{2}\vec{B}\cdot\vec{B} - A_0\left(\nabla\cdot\vec{E}\right)$$

So A_0 acts as a Lagrange multiplier which imposes Gauss' law

$$\nabla\cdot\vec{E} = 0$$

which is now a constraint on the system in which $\vec{A}$ are the physical degrees of freedom. Let's now see how to treat this system using different gauge fixing conditions.

Coulomb Gauge

In Coulomb gauge, the equation of motion for $\vec{A}$ is,

$$\partial_\mu\partial^\mu \vec{A} = 0$$

which we can solve in the usual way,

$$\vec{A} = \int\frac{d^3 p}{(2\pi)^3}\vec{\xi}(\vec{p})e^{ip\cdot x}$$

with $p_0^2 = |\vec{p}|^2$. The constraint $\nabla\cdot\vec{A} = 0$ tells us that $\vec{\xi}$ must satisfy,

$$\vec{\xi}\cdot p = 0$$

which means that $\vec{\xi}$ is perpendicular to the direction of motion $\vec{p}$. We can pick $\vec{\xi}(\vec{\ })$ to be a linear combination of two orthonormal vectors $\vec{\epsilon}_r$, $r = 1, 2$, each of which satisfies $\vec{\epsilon}_r(\vec{p})\cdot\vec{p} = 0$ and

$$\vec{\epsilon}_r(\vec{p})\cdot\vec{\epsilon}_s(\vec{p}) = \delta_{rs} \quad r, s = 1, 2.$$

These two vectors correspond to the two polarization states of the photon. It's worth pointing out that you can't consistently pick a continuous basis of polarization vectors for every value of ~p because you can't comb the hair on a sphere. But this topological fact doesn't cause any complications in computing QED scattering processes.

To quantize we turn the Poisson brackets into commutators. Naively we would write:

$$\left[A_i\left(\vec{x}\right), A_j\left(\vec{y}\right)\right] = \left[E^i\left(\vec{x}\right), E^j\left(\vec{y}\right)\right] = 0$$

$$\left[A_i\left(\vec{x}\right), E^j\left(\vec{y}\right)\right] = i\delta^j{}_i\, \delta^{(3)}\left(\vec{x} - \vec{y}\right)$$

But this can't quite be right, because it's not consistent with the constraints. We still want to have $\nabla \cdot \vec{A} = \nabla \cdot \vec{E} = 0$, now imposed on the operators. But from the commutator relations above,

$$\left[\nabla \cdot \vec{A}_{(\vec{x})}, \nabla \cdot \vec{E}\left(\vec{y}\right)\right] =_i \nabla^2\, \delta^{(3)}\left(\vec{x} - \vec{y}\right) \neq 0$$

In imposing the commutator relations $\left[A_i\left(\vec{x}\right), E^j\left(\vec{y}\right)\right] = i\delta^j{}_i\, \delta^{(3)}\left(\vec{x} - \vec{y}\right)$ we haven't correctly taken into account the constraints. In fact, this is a problem already in the classical theory, where the Poisson bracket structure is already altered. The correct Poisson bracket structure leads to an alteration of the last commutation relation,

$$[A_i\left(\vec{x}\right), E_j\left(\vec{y}\right)] = i - \left(\delta_{ij} \frac{\partial_i \partial_j}{\nabla^2}\right) \delta^{(3)}\left(\vec{x} - \vec{y}\right)$$

To see that this is now consistent with the constraints, we can rewrite the right-hand side of the commutator in momentum space,

$$\left[A_i\left(\vec{x}\right), E_j\left(\vec{y}\right)\right] = i \int \frac{d^3p}{(2\pi)^3} \left(\delta_{ij} - \frac{p_i p_j}{\left|\vec{p}\right|^2}\right) e^{i\vec{p}\cdot(\vec{x}-\vec{y})}$$

which is now consistent with the constraints, for example:

$$\left[\partial_i A_i\left(\vec{x}\right), E_j\left(\vec{y}\right)\right] = i \int \frac{d^3p}{(2\pi)^3} \left(\delta_{ij} - \frac{p_i p_j}{\left|\vec{p}\right|^2}\right) ip_i\, e^{i\vec{p}\cdot(\vec{x}-\vec{y})} = 0$$

We now write $\vec{A}$ in the usual mode expansion,

$$\vec{A}\left(\vec{x}\right) \int \frac{d^3\, p}{(2\pi)^3} \frac{1}{\sqrt{2\left|\vec{p}\right|}} \sum_{r=1}^{2} \vec{\epsilon}_r\left(\vec{p}\right)\left[a^r_{\vec{p}}\, e^{i\vec{p}\cdot\vec{x}} + a^{ar^\dagger}_{\vec{p}}\, e^{-i\vec{p}\cdot\vec{x}}\right]$$

$$\vec{E}\left(\vec{x}\right) \int \frac{d^3p}{(2\pi)^3} (-i) \sqrt{\frac{\left|\vec{p}\right|}{2}} \sum_{r=1}^{2} \vec{\epsilon}_r\left(\vec{p}\right)\left[a^r_{\vec{p}}\, e^{i\vec{p}\cdot\vec{x}} - a^{r^\dagger}_{\vec{p}}\, e^{-i\vec{p}\cdot\vec{x}}\right]$$

where, as before, the polarization vectors satisfy,

$$\vec{\epsilon}_r(\vec{p}) \cdot \vec{p} = 0 \text{ and } \vec{\epsilon}_r(\vec{p}) \cdot \vec{\epsilon}_s(\vec{p}) = \delta_{rs}$$

It is not hard to show that the commutation relations $[A_i(\vec{x}), E_j(\vec{y})] = i - \left(\delta_{ij} \frac{\partial_i \partial_j}{\nabla^2}\right)\delta^{(3)}(\vec{x} - \vec{y})$ are equivalent to the usual commutation relations for the creation and annihilation operators,

$$\left[a^r_{\vec{p}}, a^s_{\vec{q}}\right] = \left[a^{r\dagger}_{\vec{p}}, a^{s\dagger}_{\vec{q}}\right] = 0$$

$$= \left[a^r_{\vec{p}}, a^{s\dagger}_{\vec{q}}\right] = (2\pi)^3 \delta^{rs} \delta^{(3)}(\vec{p} - \vec{q})$$

where, in deriving this, we need the completeness relation for the polarization vectors,

$$\sum_{r=1}^{2} \epsilon^i_r(\vec{p}) \epsilon^j_r(\vec{p}) = \delta^{ij} - \frac{p^i p^j}{|\vec{p}|^2}$$

You can easily check that this equation is true by acting on both sides with a basis of vectors $\vec{\epsilon}_1 1(\vec{p}), \vec{\epsilon}_2(\vec{p}), \vec{p})$.

We derive the Hamiltonian by substituting $\vec{E}(\vec{x}) \int \frac{d^3p}{(2\pi)^3} (\text{-i}) \sqrt{\frac{|\vec{p}|}{2}} \sum_{r=1}^{2} \vec{\epsilon}_r(\vec{p}) \left[a^r_{\vec{p}} e^{i\vec{p}\cdot\vec{x}} - a^{r\dagger}_{\vec{p}} e^{-i\vec{p}\cdot\vec{x}}\right]$ into $= \int d^3x \frac{1}{2} \vec{E} \cdot \vec{E} + \frac{1}{2} \vec{B} \cdot \vec{B} - A_0 (\nabla \cdot \vec{E})$. The last term vanishes in Coulomb gauge. After normal ordering, and playing around with $\vec{\epsilon}_r$ r polarization vectors, we get the simple expression,

$$H = \int \frac{d^3p}{(2\pi)^3} |\vec{p}| \sum_{r=1}^{2} a^{r\dagger}_{\vec{p}} a^r_{\vec{p}}$$

The Coulomb gauge has the advantage that the physical degrees of freedom are manifest. However, we've lost all semblance of Lorentz invariance. One place where this manifests itself is in the propagator for the fields Ai(x) (in the Heisenberg picture). In Coulomb gauge the propagator reads,

$$D^{tr}_{ij}(x - y) \equiv \langle 0| TA_i(x) A_j(y) |0\rangle = \int \frac{d^4p}{(2\pi)^4} \frac{i}{p2 + i\xi} \left(\delta_{ij} - \frac{p_i p_j}{|\vec{p}|^2}\right) e^{-ip\cdot(x-y)}$$

The *tr* superscript on the propagator refers to the "transverse" part of the photon. When we turn to the interacting theory, we will have to fight to massage this propagator into something a little nicer.

Lorentz Gauge

We could try to work in a Lorentz invariant fashion by imposing the Lorentz gauge condition $\partial_\mu A^u = 0$. The equations of motion that follow from the action are then,

$$\partial_\mu \partial^\mu A^\nu = 0$$

Our approach to implementing Lorentz gauge will be a little different from the method we used in Coulomb gauge. We choose to change the theory so that $\partial_\mu \partial^\mu A^\nu = 0$ arises directly through the equations of motion. We can achieve this by taking the Lagrangian,

$$L = -\frac{1}{4} F_{\mu\nu} F^{\mu\nu} - \frac{1}{2} \left(\partial_\mu A^\mu\right)^2.$$

The equations of motion coming from this action are,

$$\partial_\mu F^{\mu\nu} + \partial^\nu \left(\partial_\mu A^\mu\right) = \partial_\mu \partial^\mu A^\nu = 0$$

In fact, we could be a little more general than this, and consider the Lagrangian,

$$\mathcal{L} = -\frac{1}{4} F_{\mu\nu} F^{\mu\nu} - \frac{1}{2\alpha} \left(\partial_\mu A^\mu\right)^2$$

with arbitrary α and reach similar conclusions. The quantization of the theory is independent of α and, rather confusingly, different choices of α are sometimes also referred to as different "gauges". We will use $\alpha = 1$, which is called "Feynman gauge". The other common choice, $\alpha = 0$, is called "Landau gauge".

Our plan will be to quantize the theory $\partial_\mu F^{\mu\nu} + \partial^\nu \left(\partial_\mu A^\mu\right) = \partial_\mu \partial^\mu A^\nu = 0$, and only later impose the constraint $\partial_\mu A^\mu = 0$ in a suitable manner on the Hilbert space of the theory. As we'll see, we will also have to deal with the residual gauge symmetry of this theory which will prove a little tricky. At first, we can proceed very easily, because both π^0 and π^i are dynamical:

$$\pi^0 = \frac{\partial \mathcal{L}}{\partial \dot{A}_0} = -\partial_\mu A^\mu$$

$$\pi^i = \frac{\partial \mathcal{L}}{\partial \dot{A}_i} = -\partial^i A^0 - A^i$$

Turning these classical fields into operators, we can simply impose the usual commutation relations,

$$\left[A_\mu(\vec{x}), A_\nu(\vec{y})\right] = \left[\pi^\mu(\vec{x}), \pi^\nu(\vec{y})\right] = 0$$

$$\left[A_\mu(\vec{x}), \pi_\nu(\vec{y})\right] = i\eta_{\mu\nu} \delta^{(3)}\left(\vec{x} - \vec{y}\right)$$

and we can make the usual expansion in terms of creation and annihilation operators and 4 polarization vectors $(\epsilon_\mu)^\lambda$, with $\lambda = 0, 1, 2, 3$.

$$A_\mu(\vec{x}) = \int \frac{d^3p}{(2\pi)^3} \frac{1}{\sqrt{2|\vec{p}|}} \sum_{\lambda=0}^{3} \epsilon^{\lambda}_{\mu}(\vec{p}) \left[a^{\lambda}_{\vec{p}} e^{i\vec{p}\vec{x}} + a^{\lambda\dagger}_{\vec{p}} e^{-i\vec{p}\vec{x}} \right]$$

$$\pi^\mu(\vec{x}) = \int \frac{d^3p}{(2\pi)^3} \sqrt{\frac{|\vec{p}|}{2}} (+i) \sum_{\lambda=0}^{3} (\epsilon^\mu)(\vec{p}) \left[a^{\lambda}_{\vec{p}} e^{i\vec{p}\vec{x}} - a^{\lambda\dagger}_{\vec{p}} e^{-i\vec{p}\vec{x}} \right]$$

Note that the momentum π^μ comes with a factor of $(+i)$, rather than the familiar $(-i)$ that we've seen so far. This can be traced to the fact that the momentum $\pi^i = \frac{\partial \mathcal{L}}{\partial \dot{A}_i} = -\partial^i A^0 - A^i$ for the classical fields takes the form $\pi^\mu = -\dot{A}^\mu + \ldots$. In the Heisenberg picture, it becomes clear that this descends to $(+i)$ in the definition of momentum.

There are now four polarization 4-vectors $\epsilon^\lambda(\vec{p})$ instead of the two polarization 3- vectors that we met in the Coulomb gauge. Of these four 4-vectors, we pick ϵ^0 to be timelike, while ϵ 1,2,3 are spacelike. We pick the normalization,

$$\epsilon^\lambda \cdot \epsilon^\lambda = \eta^{\lambda\lambda'}$$

which also means that,

$$(\epsilon_\mu)^\lambda (\epsilon \nu)^{\lambda'} \eta\lambda\lambda' = \eta_{\mu\nu}$$

The polarization vectors depend on the photon 4-momentum p = $p = (|\vec{p}|, \vec{p})$ Of the two spacelike polarizations, we will choose ϵ^1 and ϵ^2 to lie transverse to the momentum:

$$\epsilon^1 \cdot p = \epsilon^2 \cdot p = 0$$

The third vector ϵ^3 is the longitudinal polarization. For example, if the momentum lies along the x^3 direction, so $p \sim (1, 0, 0, 1)$, then

$$\epsilon^0 = \begin{pmatrix} 1 \\ 0 \\ 0 \\ 0 \end{pmatrix}, \quad \epsilon^1 = \begin{pmatrix} 0 \\ 1 \\ 0 \\ 0 \end{pmatrix}, \epsilon^2 = \begin{pmatrix} 0 \\ 0 \\ 1 \\ 0 \end{pmatrix}, \epsilon^3 = \begin{pmatrix} 0 \\ 0 \\ 0 \\ 1 \end{pmatrix}$$

For other 4-momenta, the polarization vectors are the appropriate Lorentz transformations of these vectors, since $\epsilon^1 \cdot p = \epsilon^2 \cdot p = 0$ are Lorentz invariant.

We do our usual trick, and translate the field commutation relations $\left[A_\mu(\vec{x}), \pi_\nu(\vec{y})\right] = i\eta_{\mu\nu}\delta^{(3)}(\vec{x}-\vec{y})$

into those for creation and annihilation operators We find $\left[a^{\lambda}_{\vec{p}}, a^{\lambda}_{\vec{q}}\right] = \left[a^{\lambda\dagger}_{\vec{p}}, a^{\lambda'\dagger}_{\vec{q}}\right] = 0$ and

$$\left[a^{\lambda}_{\vec{p}}, a^{\lambda'\dagger}_{\vec{q}}\right] = -\eta^{\lambda\lambda'} (2\pi)^3 \delta^{(3)}\left(\vec{p}-\vec{q}\right)$$

The minus signs here are odd to say the least. For spacelike λ = 1, 2, 3, everything looks fine,

$$= \left[a^{\lambda}_{\vec{p}}, a^{\lambda'\dagger}_{\vec{q}}\right] \delta^{\lambda\lambda'} (2\pi)^3 \delta^{(3)}\left(\vec{p}-\vec{q}\right) \quad \lambda, \lambda' = 1, 2, 3$$

But for the timelike annihilation and creation operators, we have,

$$= \left[a^{0}_{\vec{p}}, a^{0'\dagger}_{\vec{q}}\right] = - (2\pi)^3 \delta^{(3)}\left(\vec{p}-\vec{q}\right)$$

how strange this is, we take the Lorentz invariant vacuum $|0\rangle$ defined by:

$$a^{\lambda}_{\vec{p}}|0\rangle = 0$$

Then we can create one-particle states in the usual way,

$$\left|\vec{p}, \lambda\right\rangle = a^{\lambda\dagger}_{\vec{p}}|0\rangle$$

For spacelike polarization states λ = 1, 2, 3, all seems well. But for the timelike polarization λ = 0, the state $\left|\vec{p},0\right\rangle$ has negative norm,

$$\left\langle\vec{p},0\right|\left.\vec{q},0\right\rangle = \left\langle 0\right| a^{0}_{\vec{p}}, a^{0'\dagger}_{\vec{q}} a\left|0\right\rangle = -(2\pi)^3 \delta^{(3)}\left(\vec{p}-\vec{q}\right)$$

A Hilbert space with negative norm means negative probabilities which makes no sense at all. We can trace this negative norm back to the wrong sign of the kinetic term for A_0 in our original Lagrangian $L = +\frac{1}{2}\dot{A}^2 - \frac{1}{2}\dot{A}_0^2 + .$

At this point we should remember our constraint equation, $\partial_\mu A^\mu = 0$, which, until now, we've not imposed on our theory. This is going to come to our rescue. We will see that it will remove the timelike, negative norm states, and cut the physical polarizations down to two. We work in the Heisenberg picture, so that:

$$\partial_\mu A^\mu = 0$$

makes sense as an operator equation. Then we could try implementing the constraint in the quantum theory in a number of different ways. At a number of increasingly weak ways to do this:

- We could ask that $\partial_\mu A^\mu = 0$ is imposed as an equation on operators. But this can't possibly work because the commutation relations $\left[A_\mu(\vec{x}),\ \pi_\nu(\vec{y})\right] = i\eta_{\mu\nu}\delta^{(3)}(\vec{x}-\vec{y})$ won't be obeyed $\pi^0 = -\partial_\mu A^\mu$. We need some weaker condition.

- We could try to impose the condition on the Hilbert space instead of directly on the operators. After all, that's where the trouble lies. We could imagine that there's some way to split the Hilbert space up into good states $|\Psi i\rangle$ and bad states that somehow decouple from the system,

$$\partial_\mu A^\mu |\Psi i\rangle = 0$$

on all good, physical states $|\Psi\rangle$. But this can't work either! Again, the condition is too strong. For example, suppose we decompose $A_\mu(x) = A^+_\mu(x) + A^-_\mu(x)$ with,

$$A^+_\mu(x) = \int \frac{d^3p}{(2\pi)^3} 1 \frac{1}{\sqrt{2|\vec{p}|}} \sum_{\lambda=0}^{3} \in^\lambda_\mu a^\lambda_p e^{-ip\cdot x}$$

$$A^+_\mu(x) = \int \frac{d^3p}{(2\pi)^3} 1 \frac{1}{\sqrt{2|\vec{p}|}} \sum_{\lambda=0}^{3} \in^\lambda_\mu a^{\lambda\dagger}_p e^{+ip\cdot x}$$

Then, on the vacuum $A^+_\mu |0\rangle = 0$ automatically, but $\partial^\mu A^-_\mu |0\rangle \neq 0$. So not even the vacuum is a physical state if we use $\partial_\mu A^\mu |\Psi i\rangle = 0$ as our constraint.

- Our final attempt will be the correct one. In order to keep the vacuum as a good.

physical state, we can ask that physical states $|\Psi\rangle$ are defined by:

$$\partial^\mu A^+_\mu |\Psi\rangle = 0$$

This ensures that,

$$\langle \Psi' | \partial^\mu A^\mu | \Psi \rangle = 0$$

so that the operator $\partial_\mu A^\mu$ µ has vanishing matrix elements between physical states. Equation $\partial^\mu A^+_\mu |\Psi\rangle = 0$ is known as the Gupta-Bleuler condition. The linearity of the constraint means that the physical states $|\Psi i\rangle$ span a physical Hilbert space $H_{phys.}$

So what does the physical Hilbert space H_{phys} look like? And, in particular, have werid ourselves of those nasty negative norm states so that H_{phys} has a positive definite inner product defined on it? The answer is actually no, but almost.

Let's consider a basis of states for the Fock space. We can decompose any element of this basis as $|\Psi\rangle = |\psi T\rangle\,|\varphi\rangle$ where $|\psi_T\rangle$ i contains only transverse photons, created by, $a^{1,2\dagger}_{\vec{p}}$, while $|\varphi\rangle$ contains the timelike photons created by $a^{0\dagger}_{\vec{p}}$ and longitudinal photons created by $a^{3\dagger}_{\vec{p}}$. The Gupta-Bleuler condition $\partial^\mu A^+_\mu|\Psi\rangle = 0$ requires,

$$\left(a^3_{\vec{p}} - a^0_{\vec{p}}\right)|\phi\rangle = 0$$

This means that the physical states must contain combinations of timelike and longitudinal photons. Whenever the state contains a timelike photon of momentum $\vec{p}$, it must also contain a longitudinal photon with the same momentum. In general $|\phi\rangle$ will be a linear combination of states $|\phi_n\rangle$ containing n pairs of timelike and longitudinal photons, which we can write as:

$$|\phi\rangle = \sum_{n=0}^{\infty} C_n\,|\phi_n\rangle$$

where $|\phi 0\rangle = |0\rangle$ is simply the vacuum. It's not hard to show that although the condition $\left(a^3_{\vec{p}} - a^0_{\vec{p}}\right)|\phi\rangle = 0$ does indeed decouple the negative norm states, all the remaining states involving timelike and longitudinal photons have zero norm,

$$\langle\phi_m\,|\,\phi_n\rangle = \delta_{n0}\delta_{m0}$$

This means that the inner product on H_{phys} is positive semi-definite. Which is an improvement. But we still need to deal with all these zero norm states.

The way we cope with the zero norm states. is to treat them as gauge equivalent to the vacuum. Two states that differ only in their timelike and longitudinal photon content, $|\phi n\rangle$ with $n \geq 1$ are said to be physically equivalent. We can think of the gauge symmetry of the classical theory as descending to the Hilbert space of the quantum theory. Of course, we can't just stipulate that two states are physically identical unless they give the same expectation value for all physical observables. We can check that this is true for the Hamiltonian, which can be easily computed to be,

$$H = \int \frac{d^3p}{(2\pi)^3}\,|\vec{p}|\left(\sum_{i=1}^{3} a^{i\dagger}_{\vec{p}}\, a^{i}_{\vec{p}} - a^{0\dagger}_{\vec{p}}\, a^{0}_{\vec{p}}\right)$$

But the condition $\left(a^3_{\vec{p}} - a^0_{\vec{p}}\right)|\phi\rangle = 0$ ensures that $\langle\Psi|\,a^{3\dagger}_{\vec{p}}\, a^{3}_{\vec{p}}\,|\Psi i\rangle = \langle\Psi|\,a^{0\dagger}_{\vec{p}}\, a^{0}_{\vec{p}}\,|\Psi\rangle$ so that the contributions from the timelike and longitudinal photons cancel amongst themselves in the Hamiltonian. This also renders the Hamiltonian positive definite, leaving us just with the contribution from the transverse photons as we would expect.

In general, one can show that the expectation values of all gauge invariant operators evaluated on physical states are independent of the coefficients C_n in $|\phi\rangle = \sum_{n=0}^{\infty} C_n\,|\phi_n\rangle$.

Propagators

Finally, it's a simple matter to compute the propagator in Lorentz gauge. It is given by:

$$\langle 0|\, T A_\mu(x) A_\nu(y) |0\rangle = \int \frac{d^4p}{(2\pi)^4} \frac{-i\eta_{\mu\nu}}{p^2 + i\epsilon} e^{-ip\cdot(x-y)}$$

This is a lot nicer than the propagator we found in Coulomb gauge: in particular, it's Lorentz invariant. We could also return to the Lagrangian $\mathcal{L} = -\frac{1}{4}F_{\mu\nu}F^{\mu\nu} - \frac{1}{2\alpha}(\partial_\mu A^\mu)^2$. Had we pushed through the calculation with arbitrary coefficient α, we would find the propagator,

$$\langle 0|\, T A_\mu(x) A_\nu(y) |0\rangle = \int \frac{d^4p}{(2\pi)^4} \frac{-i}{p^2+i\epsilon}\left(\eta_{\mu\nu} + (\alpha-1)\frac{p_\mu p_\nu}{p^2}\right) e^{-ip\cdot(x-y)}$$

Coupling to Matter

Let's now build an interacting theory of light and matter. We want to write down a Lagrangian which couples Aμ to some matter fields, either scalars or spinors. For example, we could write something like,

$$\mathcal{L} = -\frac{1}{4}F_{\mu\nu}F^{\mu\nu} - j^\mu A_\mu$$

where j^μ is some function of the matter fields. The equations of motion read,

$$\partial_\mu F^{\mu\nu} = j^\nu$$

so, for consistency, we require:

$$\partial_\mu j^\mu = 0$$

In other words j^μ must be a conserved current. But we've got lots of those! Let's look at how we can couple two of them to electromagnetism.

Coupling to Fermions

The Dirac Lagrangian,

$$\mathcal{L} = \bar{\psi}\,(i\partial\!\!\!/ - m)\psi$$

has an internal symmetry $\psi \to e^{-i\alpha\psi}$ and $\bar{\psi} \to e^{+i\alpha}\,\bar{\psi}$ with $\alpha \in R$. This gives rise to the conserved current $j_V^\mu = \bar{\psi}\,\gamma^\mu\psi$. So we could look at the theory of electromagnetism coupled to fermions, with the Lagrangian,

$$\mathcal{L} = -\frac{1}{4}F_{\mu\nu}F^{\mu\nu} + \bar{\psi}(i\partial\!\!\!/ - m)\psi - e\,\bar{\psi}\gamma^\mu A_\mu\psi$$

where we've introduced a coupling constant *e*. For the free Maxwell theory, we have seen that the existence of a gauge symmetry was crucial in order to cut down the physical degrees of freedom to the requisite 2. Does our interacting theory above still have a gauge symmetry? Let's rewrite the Lagrangian as:

$$\mathcal{L} = -\frac{1}{4} F_{\mu\nu}F^{\mu\nu} + \bar{\psi}\left(i\not{D} - m\right)\psi$$

where $D_\mu\psi = \partial_\mu\psi + ieA_\mu\psi$ is called the covariant derivative. This Lagrangian is invariant under gauge transformations which act as:

$$A_\mu \to A_\mu + \partial_\mu\lambda \; and \; \psi \to e^{-ie\lambda\psi}$$

for an arbitrary function $\lambda(x)$. The tricky term is the derivative acting on ψ, since this will also hit the $e^{-ie\lambda}$ piece after the transformation. To see that all is well, let's look at how the covariant derivative transforms.

We have,

$$D_\mu\psi = \partial_\mu\psi + ieA_\mu\psi$$

$$\to \partial_\mu\left(e^{-ie\lambda}\psi\right) + ie\left(A_\mu + \partial_\mu\lambda\right)\left(e^{-ie\lambda}\psi\right)$$

$$= e^{-ie\lambda} D_\mu\psi$$

so the covariant derivative has the nice property that it merely picks up a phase under the gauge transformation, with the derivative of $e^{-ie\lambda}$ cancelling the transformation of the gauge field. This ensures that the whole Lagrangian is invariant, since $\bar{\psi} \to e^{+ie\lambda(x)}\bar{\psi}$.

Electric Charge

The coupling e has the interpretation of the electric charge of the ψ particle. This follows from the equations of motion of classical electromagnetism ∂μF $\partial_\mu F^{\mu\nu} = j^\nu$ we know that the j^0 component is the charge density. We therefore have the total charge Q given by,

$$Q = e\int d^3x\, \bar{\psi}(\vec{x})\gamma^0\psi(\vec{x})$$

After treating this as a quantum equation, we have,

$$Q = e\int \frac{d^3p}{(2\pi)^3} \sum_{s=1}^{2}\left(b_{\vec{p}}^{s\dagger} b_{\vec{p}}^{s} - c_{\vec{p}}^{s\dagger} c_{\vec{p}}^{s}\right)$$

which is the number of particles, minus the number of antiparticles. Note that the particle and the anti-particle are required by the formalism to have opposite electric charge. For QED, the theory of light interacting with electrons, the electric charge is usually written in terms of the dimensionless ratio α, known as the fine structure constant,

$$\alpha = \frac{e^2}{4\pi\hbar c} \approx \frac{1}{137}$$

Setting $\hbar = c = 1$, we have $e = \sqrt{4\pi\alpha} \approx 0.3$.

There's a small subtlety here that's worth elaborating on. I stressed that there's a radical difference between the interpretation of a global symmetry and a gauge symmetry. The former takes you from one physical state to another with the same properties and results in a conserved current through Noether's theorem. The latter is a redundancy in our description of the system. Yet in electromagnetism, the gauge symmetry $\psi \to e^{+ie\lambda(x)}\psi$ seems to lead to a conservation law, namely the conservation of electric charge. This is because among the infinite number of gauge symmetries parameterized by a function $\lambda(x)$ there is also a single global symmetry: that with $\lambda(x)$ = constant.

This is a true symmetry of the system, meaning that it takes us to another physical state. More generally, the subset of global symmetries from among the gauge symmetries are those for which $\lambda(x) \to \alpha =$ constant as $x \to \infty$ These take us from one physical state to another.

Finally, let's check that the 4 × 4 matrix C really deserves the name "charge conjugation matrix". If we take the complex conjugation of the Dirac equation, we have

$$\left(i\gamma^\mu \partial^\mu - e\gamma^\mu A_\mu - m\right)\psi = 0 \Rightarrow \left(-i\left(\gamma^\mu\right)^\star \partial_\mu - e\left(\gamma^\mu\right)^\star A_\mu - m\right)\psi^\star = 0$$

Now using the defining equation $C^\dagger \gamma^\mu C = -\left(\gamma^\mu\right)^\star$ and the definition $\psi^{(c)} = C\psi^\star$ we see that the charge conjugate spinor $\psi^{(c)}$ satisfies

$$\left(i\gamma^\mu \partial_\mu + e\gamma^\mu A_\mu - m\right)\psi^{(c)} = 0$$

So we see that the charge conjugate spinor $\psi^{(c)}$ satisfies the Dirac equation, but with charge – *e*. instead of + *e*.

Coupling to Scalars

For a real scalar field, we have no suitable conserved current. This means that we can' couple a real scalar field to a gauge field Let's now consider a complex scalar field φ. We have a symmetry $\varphi \to e^{-i\alpha}\varphi$. We could try to couple the associated current to the gauge field,

$$\mathcal{L}_{int} = -i\left(\left(\partial_\mu \varphi^\star\right)\varphi \;-\; \varphi^\star \partial_\mu \phi\right)A^\mu$$

But this doesn't work because:

- The theory is no longer gauge invariant.
- The current j^μ that we coupled to A^μ depends on $\partial_\mu \varphi$. This means that if we try to compute

the current associated to the symmetry, it will now pick up a contribution from the $j^\mu A_\mu$ term. So the whole procedure wasn't consistent.

We solve both of these problems simultaneously by remembering the covariant derivative. In this scalar theory, the combination,

$$D_\mu \varphi = \partial_\mu \varphi + ieA_\mu \varphi$$

again transforms as $D_\mu\varphi \to e^{-ie\lambda} D_\mu \varphi$ under a gauge transformation $A_\mu \to A_\mu + \partial_\mu \lambda$ and $\varphi \to e^{-ie\lambda}\varphi$ This means that we can construct a gauge invariant action for a charged scalar field coupled to a photon simply by promoting all derivatives to covariant derivatives,

$$\mathcal{L} = -\frac{1}{4} F_{\mu\nu}F^{\mu\nu} + D_\mu\varphi^\star D^\mu \varphi - m^2 |\varphi|^2$$

In general, this trick works for any theory. If we have a $U(1)$ symmetry that we wish to couple to a gauge field, we may do so by replacing all derivatives by suitable covariant derivatives. This procedure is known as minimal coupling.

QED

Let's now work out the Feynman rules for the full theory of quantum electrodynamics (QED) – the theory of electrons interacting with light. The Lagrangian is,

$$\mathcal{L} = -\frac{1}{4} F_{\mu\nu}F^{\mu\nu} + \bar{\psi}\left(i\not{D} - m\right)\psi$$

where $D_\mu = \partial_\mu + ieA_\mu$.

The route we take now depends on the gauge choice. If, however, we worked in Coulomb gauge, then we still have a bit of work in front of us in order to massage the photon propagator into something Lorentz invariant.

In Coulomb gauge we can rewrite the Maxwell part of the Lagrangian as:

$$L_{Maxwell} = \int d^3x \; \frac{1}{2}\vec{E}^2 - \frac{1}{2}\vec{B}^2$$

$$= \int d^3x \; \frac{1}{2}\left(\vec{A} + \nabla A_0\right)^2 - \frac{1}{2}\vec{B}^2$$

$$= \int d^3x \frac{1}{2}\dot{\vec{A}}^2 + \frac{1}{2}(\nabla A_0)^2 - \frac{1}{2}\vec{B}^2$$

where the cross-term has vanished using $\nabla \cdot \vec{A} = 0$. After integrating the second term by parts and inserting the equation for A_0, we have,

$$L_{Maxwell} = \int d^3x \left[\frac{1}{2}\dot{\vec{A}}^2 - \frac{1}{2}\vec{B}^2 + \frac{e^2}{2}\int d^{3x'} \frac{jo(\vec{x})jo(\vec{x}')}{4\pi|\vec{x} - \vec{x}'|} \right]$$

We find ourselves with a nonlocal term in the action. It appears here as an artifact of working in Coulomb gauge: it does not mean that the theory of QED is nonlocal. For example, it wouldn't appear if we worked in Lorentz gauge. We now compute the Hamiltonian. Changing notation slightly from previous chapters, we have the conjugate momenta,

$$\vec{\Pi} = \frac{\partial \mathcal{L}}{\partial \dot{\vec{A}}} = \dot{\vec{A}}$$

$$\pi_\psi = \frac{\partial \mathcal{L}}{\partial \dot{\psi}} = i\psi^\dagger$$

which gives us the Hamiltonian,

$$H = \int d^3x \left[\frac{1}{2}\dot{\vec{A}}^2 + \frac{1}{2}\vec{B}^2 + \bar{\psi}\left(-i\gamma^i \partial_i + m\right)\psi - e\vec{j}\cdot\vec{A} + \frac{e^2}{2}\int d^3x' \,\frac{j_0(\vec{x})\, j_0(\vec{x}')}{4\pi\left|\vec{x} - \vec{x}'\right|} \right]$$

where $\vec{j} = \bar{\psi}\vec{\gamma}\psi$ and $j^0 = \bar{\psi}\gamma^0\psi$.

Naive Feynman Rules

We want to determine the Feynman rules for this theory. For fermions, the rules are the same as those given. The new pieces are:

- We denote the photon by a wavy line. Each end of the line comes with an $i, j = 1, 2, 3$ index telling us the component of $\vec{A}$. We calculated the transverse photon propagator in

  $D^{tr}_{ij}(x - y) \equiv \langle 0 | TA_i(x) A_j(y) | 0 \rangle = \int \frac{d^4p}{(2\pi)^4} \frac{i}{p2 + i\xi}\left(\delta_{ij} - \frac{p_i p_j}{|\vec{p}|^2}\right) e^{-ip\cdot(x-y)}$: it is 

  and contributes $D^{tr}_{ij} = \frac{i}{p^2 + i\epsilon}\left(\delta_{ij} - \frac{p_i p_j}{|\vec{p}|^2}\right)$

- The vertex 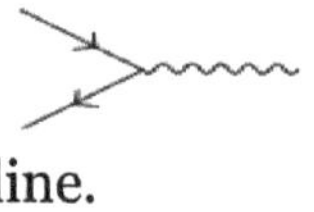 contributes $-ie\gamma^i$. The index on γ^i contracts with the index on the photon line.

- The non-local interaction which, in position space, is given by  contributes a factor of $\frac{i(e\gamma^0)^2\,\delta(x^0 - y^0)}{4\pi|\vec{x} - \vec{y}|}$

These Feynman rules are rather messy. This is the price we've paid for working in Coulomb gauge. We'll now show that we can massage these expressions into something much more simple and Lorentz invariant. Let's start with the offending instantaneous interaction. Since it comes from the A_0 component of the gauge field, we could try to redefine the propagator to include a D_{00} piece which will capture this term. In fact, it fits quite nicely in this form: if we look in momentum space, we have,

$$\frac{\delta\left(x^0 - y^0\right)}{4\pi\left|\vec{x} - \vec{y}\right|} = \int \frac{d^4 p}{(2\pi)^4} \frac{e^{ip\cdot(x-y)}}{\left|\vec{p}\right|^2}$$

so we can combine the non-local interaction with the transverse photon propagator by defining a new photon propagator,

$$D_{\mu\nu}(p) = \begin{cases} +\frac{i}{\left|\vec{p}\right|^2} & \mu,\nu = 0 \\ \frac{i}{p^2 + i\in}\left(\delta_{ij} - \frac{p_i p_j}{\left|\vec{p}\right|^2}\right) & \mu = i \neq 0, \nu = j \neq 0 \\ 0 & \end{cases}$$

With this propagator, the wavy photon line now carries a μ, $\nu = 0, 1, 2, 3$ index, with the extra $\mu = 0$ component taking care of the instantaneous interaction. We now need to change our vertex slightly: the $-ie\gamma^i$ above gets replaced by $-ie\gamma^u$ which correctly accounts for the $\left(e\gamma^0\right)^2$ piece in the instantaneous interaction.

The D_{00} piece of the propagator doesn't look a whole lot different from the transverse photon propagator. But wouldn't it be nice if they were both part of something more symmetric! In fact, they are. We have the following:

Claim: We can replace the propagator Dμν (p) with the simpler, Lorentz invariant propagator,

$$D_{\mu\nu}(p) = -i\frac{\eta_{\mu\nu}}{p^2}$$

Proof: There is a general proof using current conservation. Here we'll be more pedestrian and show that we can do this for certain Feynman diagrams. In particular, we focus on a particular tree-level diagram that contributes to $e^- e^- \to e^- e^-$ scattering,

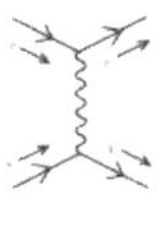

$$e^2 \left[\bar{u}(p')\gamma^\mu u(p)\right] D_{\mu\nu}(k) \left[\bar{u}(q')\gamma^\nu u(q)\right]$$

where $k = p - p' = q' - q$. Recall that u $u(\vec{p})$ satisfies the equation,

$$\left(\not{p} - m\right)u\left(\vec{p}\right) = 0$$

Let's define the spinor contractions $\alpha^\mu = \bar{u}\left(\vec{p}\,'\right)\gamma^\mu u\left(\vec{p}\right)$ *and* $\beta^\nu = \bar{u}\left(\vec{q}\,'\right)\gamma^\nu u\left(\vec{q}\right)$. Then since $k = p - p' = q' - q$, we have,

$$k_\mu \alpha^\mu = \bar{u}\left(\vec{p}\,'\right)\left(\not{p} - \not{p}\,'\right)u\left(\vec{p}\right) = \bar{u}\left(\vec{p}\,'\right)(m - m)u\left(\vec{p}\right) = 0$$

and, similarly, $k_\nu \beta^\nu = 0$. Using this fact, the diagram can be written as:

$$\begin{aligned}
\alpha^\mu D_{\mu\nu}\beta^\nu &= i\left(\frac{\vec{\alpha}\cdot\vec{\beta}}{k^2} - \frac{(\vec{\alpha}\cdot\vec{k})(\vec{\beta}\cdot\vec{k})}{k^2\,|\vec{k}|^2} + \frac{\alpha^0\beta^0}{|\vec{k}|^2}\right) \\
&= i\left(\frac{\vec{\alpha}\cdot\vec{\beta}}{k^2} - \frac{k_0^2\,\alpha_0\beta_0}{k^2\,|\vec{k}|^2} + \frac{\alpha_0\beta_0}{|\vec{k}|^2}\right) \\
&= i\frac{\vec{\alpha}\cdot\vec{\beta}}{k^2} - \frac{1}{k^2\,|\vec{k}|^2}\left(k_0^2 - k^2\right)\alpha_0\beta_0 \\
&= -\frac{i}{k^2}\alpha\cdot\beta = \alpha^\mu\left(\frac{i\eta_{\mu\nu}}{k^2}\right) - \beta^\nu
\end{aligned}$$

which is the claimed result. You can similarly check that the same substitution is legal in the diagram,

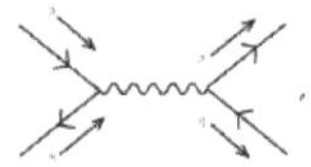
$$\sim e^2\left[\bar{v}(\vec{q})\gamma^\mu u(\vec{p})\right] D_{\mu\nu}(k)\left[\bar{u}(\vec{p}')\gamma^\nu v(\vec{q}')\right]$$

n fact, although we won't show it here, it's a general fact that in every Feynman diagram we may use the very nice, Lorentz invariant propagator $D_{\mu\nu} = -i\eta_{\mu\nu}/p^2$.

This is the propagator we found when quantizing in Lorentz gauge (using the Feynman gauge parameter). In general, quantizing the Lagrangian $\mathcal{L} = -\frac{1}{4}F_{\mu\nu}F^{\mu\nu} - \frac{1}{2\alpha}\left(\partial_\mu A^\mu\right)^2$ in Lorentz gauge, we have the propagator.

$$D_{\mu\nu} = -\frac{i}{p^2}\left(\eta_{\mu\nu} + (\alpha - 1)\frac{p_\mu p_\nu}{p^2}\right)$$

Using similar arguments to those given above, you can show that the $p_\mu p_\nu/p^2$ term cancels in all diagrams. For example, in the following diagrams the $p_\mu p_\nu$ piece of the propagator contributes as

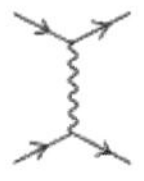
$$\sim \bar{u}(p')\gamma^\mu u(p)k_\mu = \bar{u}(p')(\not{p} - \not{p}')u(p) = 0$$

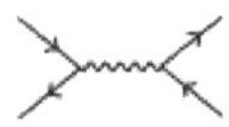
$$\sim \bar{u}(p)\gamma^\mu u(q)k_\mu = \bar{u}(p)(\not{p} - \not{p}')u(q) = 0$$

Feynman Rules

Finally, we have the Feynman rules for QED. For vertices and internal lines, we write:

- Vertex: 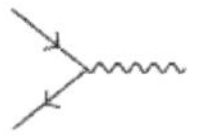 $-ie\gamma^\mu$
- Photon Propagator:  $-\dfrac{i\eta_{\mu\nu}}{p^2 + i\epsilon}$

- Fermion Propagator:  $\dfrac{i(\not{p}+m)}{p^2-m^2+i\epsilon}$

For external lines in the diagram, we attach

- Photons: We add a polarization vector $\epsilon_{in}^{\mu} / \epsilon_{out}^{\mu}$ for incoming/outgoing photons. In Coulomb gauge $\epsilon^0 = 0$ *and* $\vec{\epsilon}.\vec{p} = 0$
- Fermions: We add a spinor $u^r(\vec{p}) / \bar{u}^r(\vec{p})$ for incoming/outgoing fermions. We add a spinor $\bar{v}^r(\vec{p}) / v^r(\vec{p})$ for incoming/outgoing anti-fermions.

Charged Scalars

The interaction terms in the Lagrangian for charged scalars come from the covariant derivative terms,

$$\mathcal{L} = D_\mu\psi^\dagger D^\mu\psi = \partial_\mu\psi^\dagger\partial^\mu\psi - ieA_\mu\left(\psi^\dagger\partial^\mu\psi - \psi\partial^\mu\psi^\dagger\right) + e^2A_\mu A^\mu\psi^\dagger\psi$$

This gives rise to two interaction vertices. But the cubic vertex is something we haven't seen before: it contains kinetic terms. How do these appear in the Feynman rules? After a Fourier transform, the derivative term means that the interaction is stronger for fermions with higher momentum, so we include a momentum factor in the Feynman rule. There is also a second, "seagull" graph. The two Feynman rules are,

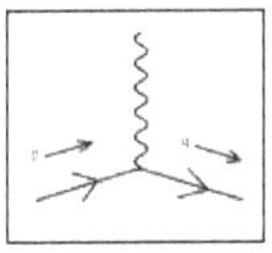 $-ie(p+q)_\mu$ and

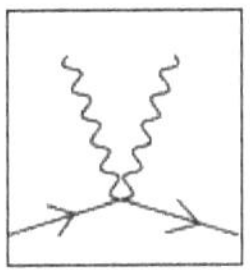 $+\,2ie^2\,\eta_{\mu\nu}$

The factor of two in the seagull diagram arises because of the two identical particles appearing in the vertex.

Scattering in QED

Let's now calculate some amplitudes for various processes in quantum electrodynamics, with a photon coupled to a single fermion.

Electron Scattering

Electron scattering $e^-e^- \to e^-e^-$ is described by the two leading order Feynman diagrams, given by

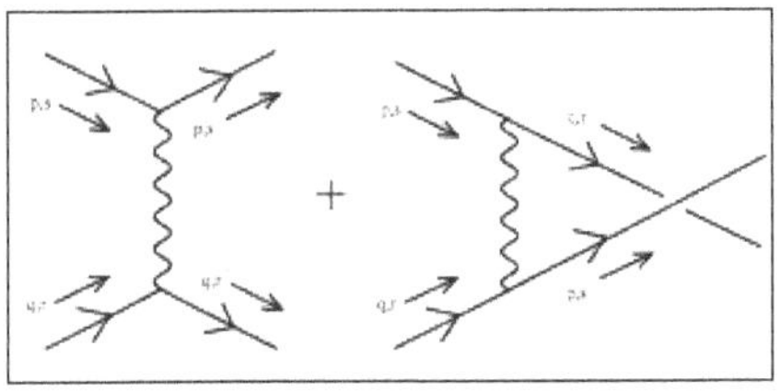

$$= -i(-ie)^2\left(\frac{\left[\bar{u}^{s'}(\vec{p}')\gamma^{\mu}u^{s}(\vec{p})\right]\left[\bar{u}^{r'}(\vec{q}')\gamma_{\mu}u^{r}(\vec{q})\right]}{(p'-p)^2}\right)-\frac{\left[\bar{u}^{s'}(\vec{p}')\gamma^{\mu}u^{r}(\vec{q})\right]\left[\bar{u}^{r'}(\vec{q}')\gamma_{\mu}u^{s}(\vec{p})\right]}{(p-q')^2}$$

The overall −i comes from the $-i\eta_{\mu\nu}$ in the propagator, which contract the indices on the γ-matrices (remember that it's really positive for $\mu,\nu = 1, 2, 3$).

Electron Positron Annihilation

Let's now look $e^-e^+ \rightarrow 2\gamma$, two gamma rays. The two lowest order Feynman diagrams are,

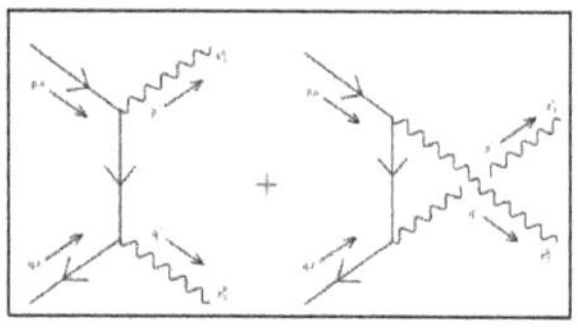

$$= -i(-ie)^2\,\bar{v}^{r}(\vec{q})\left(\frac{\gamma_{\mu}(\not{p}-\not{p}'+m)\gamma\upsilon}{(p-p')^2-m^2}+\frac{\gamma_{\upsilon}(\not{p}-\not{q}'+m)\gamma_{\mu}}{(p-q')^2-m^2}\right)u^{s}(\vec{p})\,\epsilon_1^{\upsilon}(\vec{p}')\,\epsilon_2^{\mu}(\vec{q}')$$

Electron Positron Scattering

For $e^-e^+ \rightarrow e^-e^+$ scattering (sometimes known as Bhabha scattering) the two lowest order Feynman diagrams are,

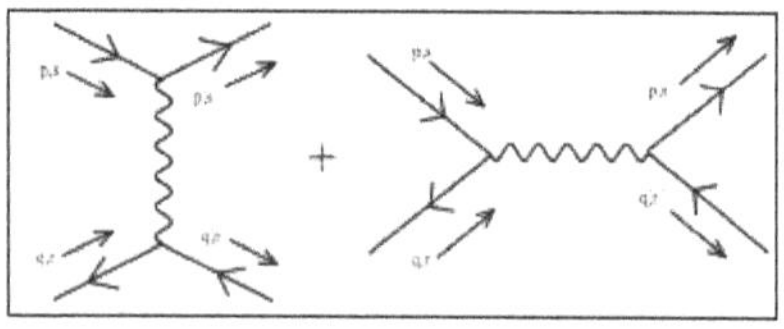

$$=\frac{-i(-ie)^2 - \left[\bar{u}^{s'}(\vec{p}')\gamma^{\mu}u^{s}(\vec{p})\right]\left[\bar{v}^{r}(\vec{q})\gamma_{\mu}v^{r'}(\vec{q}')\right]}{(p-p')^2}$$

$$+\frac{\left[\bar{v}^{r'}(\vec{q}')\gamma^{\mu}u^{s}(\vec{p})\right]\left[\bar{u}^{s'}(\vec{p})\gamma_{\mu}v^{r'}(\vec{q}')\right]}{(p+q)^2}$$

Compton Scattering

The scattering of photons (in particular x-rays) off electrons $e^-\gamma \rightarrow e^-\gamma$ is known as Compton scattering. Historically, the change in wavelength of the photon in the scattering process was one of the conclusive pieces of evidence that light could behave as a particle. The amplitude is given by,

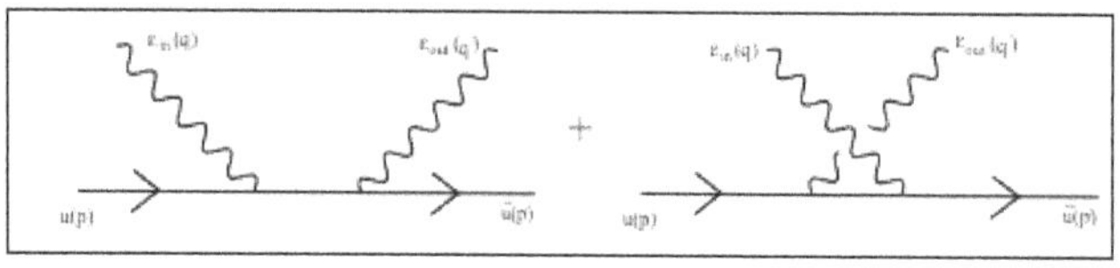

$$= i(-ie)^2 \bar{u}^{r}{}'(\vec{p}') \left(\frac{\gamma_\mu(\not{p}+\not{q}+m)\gamma\nu}{(p+q)^2 - m^2} + \frac{\gamma\nu(\not{p}-\not{q}'+m)\gamma\mu}{(p-q')^2 - m^2} \right) u^s(\vec{p}) \in^{\nu}_{in} \in^{\mu}_{out}$$

This amplitude vanishes for longitudinal photons. For example, suppose $\epsilon_{in} \sim q$. Then, using momentum conservation $p+q = p'+q$ we may write the amplitude as:

$$iA = i(-ie)^2 \bar{u}^{r}{}'(\vec{p}') \left(\frac{\gamma_{out}(\not{p}+\not{q}+m)\not{q}}{(p+q)^2 - m^2} + \frac{\not{q}(\not{p}'-\not{q}'+m)\gamma_{out}}{(p-q')^2 - m^2} \right) u^s(\vec{p})$$

$$= i(-ie)^2 \bar{u}^{r'}(\vec{p}') \not{\epsilon}_{out} u^s(\vec{p}) \left(\frac{2p \cdot q}{(p+q)^2 - m^2} + \frac{2p' \cdot q}{(p'-q)^2 - m^2} \right)$$

where, in going to the second line, we've performed some γ-matrix manipulations, and invoked the fact that q is null, so $\not{q}\not{q} = 0$, together with the spinor equations $(\not{p} - m)u(\vec{p})$ and $u(\vec{p}')(\not{p}' - m) = 0$. We now recall the fact that q is a null vector, while $p^2 = (p')^2 = m^2$ since the external legs are on mass-shell. This means that the two denominators in the amplitude read $(p+q)^2 - m^2 = 2p{\cdot}q$ and $(p'-q) - m^2 = -2p'{\cdot}q$. This ensures that the combined amplitude vanishes for longitudinal photons as promised. A similar result holds when $\not{\epsilon}_{out} \sim q'$.

Photon Scattering

In QED, photons no longer pass through each other unimpeded. At one-loop, there is a diagram which leads to photon scattering. Although naively logarithmically divergent, the diagram is actually rendered finite by gauge invariance.

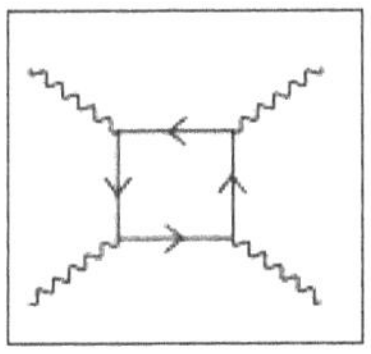

Adding Muons

Adding a second fermion into the mix, which we could identify as a muon, new processes become possible. For example, we can now have processes such as $e^-\mu^- \to e^-\mu^-$ scattering, and e^+e^- annihilation into a muon anti-muon pair. Using our standard notation of p and q for incoming momenta, and p' and q' for outgoing momenta, we have the amplitudes given by,

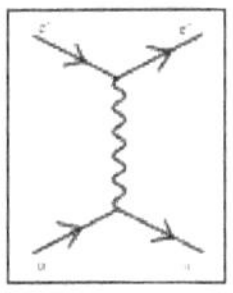 $\sim \dfrac{1}{(p-p')^2}$

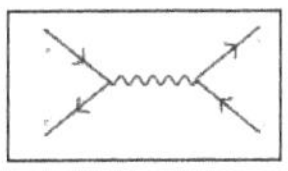 $\sim \dfrac{1}{(p+q)^2}$

Coulomb Potential

We have,

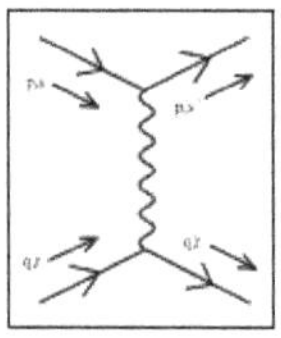

$$= -i(-ie)^2 \frac{\left[\bar{u}(\vec{p}')\gamma^\mu u(\vec{p})\right]\left[\bar{u}(\vec{q}')\gamma_\mu u(\vec{q})\right]}{(p'-p)2}$$

The non-relativistic limit of the spinor is $u(p) \to \sqrt{m}\begin{pmatrix}\xi \\ \xi\end{pmatrix}$.

This means that the γ^0 piece of the interaction gives a term $\bar{u}^s(\vec{p})\gamma^0 u^r(\vec{q}) \to 2m\delta^{rs}$, while the spatial $\gamma^i, i=1,2,3$ pieces vanish in the non-relativistic limit: $\bar{u}^s(\vec{p})\gamma^i u^r(\vec{q}) \to 0$. Comparing the scattering amplitude in this limit to that of non-relativistic quantum mechanics, we have the effective potential between two electrons given by:

$$U(\vec{r}) = +e^2 \int \frac{d^3p}{(2\pi)^3} \frac{e^{i\vec{p}\cdot\vec{r}}}{|\vec{p}|^2} = +\frac{e^2}{4\pi r}$$

We find the familiar repulsive Coulomb potential. We can trace the minus sign that gives a repulsive potential to the fact that only the A0 component of the intermediate propagator $\sim -i\eta_{\mu\nu}$ contributes in the non-relativistic limit. For $e^-e^+ \to e^-e^+$ scattering, the amplitude is:

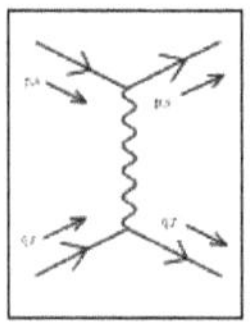

$$= \frac{+i(-ie)^2\left[\bar{u}(\vec{p})\gamma^\mu u(\vec{p})\right]\left[\bar{v}(\vec{q})\gamma_\mu v(\vec{q}')\right]}{(p'-p)^2}$$

The overall + sign comes from treating the fermions correctly: we saw the same minus sign when studying scattering in Yukawa theory. The difference now comes from looking at the non-relativistic limit. We have $\bar{v}\gamma^0 v \to 2m$, giving us the potential between opposite charges,

$$U(\vec{r}) = -e^2 \int \frac{d^3p}{(2\pi)^3} \frac{e^{i\vec{p}\cdot\vec{r}}}{|\vec{p}|^2} = -\frac{e^2}{4\pi r}$$

Reassuringly, we find an attractive force between an electron and positron. The difference from the calculation of the Yukawa force comes again from the zeroth component of the gauge field, this time in the guise of the γ^0 sandwiched between $\bar{v}\gamma^0 v \to 2m$, rather than the $\bar{v}v \to -2m$ that we saw in the Yukawa case.

Coulomb Potential for Scalars

There are many minus signs in the above calculation which somewhat obscure the crucial one which gives rise to the repulsive force. A careful study reveals the offending sign to be that which sits in front of the A0 piece of the photon propagator $-i\eta_{\mu\nu}/p^2$ Note that with our signature $(+---)$, the propagating Ai have the correct sign, while Ao comes with the wrong sign. This is simpler to see in the case of scalar QED, where we don't have to worry about the gamma matrices. From the Feynman rules, we have the non-relativistic limit of scalar e^-e^- scattering,

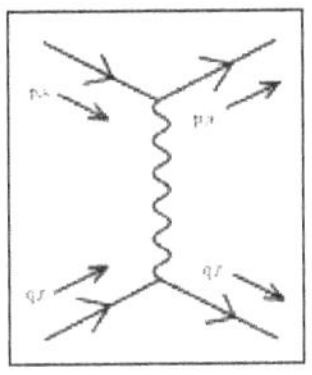

$$= -i\eta_{\mu\nu}\left(-ie\right)^2 \frac{\left(p+p'\right)^\mu \left(q+q'\right)_\nu}{\left(p'-p\right)^2} \to -i\left(-ie\right)^2 \left(2m\right)^2 - \left(\vec{p}-\vec{p}'\right)^2$$

where the non-relativistic limit in the numerator involves $(p+p')\cdot(q+q') \approx (p+p')^0 (q+q')_0 \approx (2m)^2$ and is responsible for selecting the A_0 part of the photon propagator rather than the A_i piece. This shows that the Coulomb potential for spin 0 particles of the same charge is again repulsive, just as it is for fermions. For e^-e^+ scattering, the amplitude picks up an extra minus sign because the arrows on the legs of the Feynman rules are correlated with the momentum arrows. Flipping the arrows on one pair of legs in the amplitude introduces the relevant minus sign to ensure that the non-relativistic potential between e^-e^+ is attractive as expected.

The Dirac Equation

In particle physics, the Dirac equation is a relativistic wave equation derived by British physicist Paul Dirac in 1928. In its free form, or including electromagnetic interactions, it describes all spin-1/2 massive particles such as electrons and quarks for which parity is a symmetry. It is consistent with both the principles of quantum mechanics and the theory of special relativity, and was the first theory to account fully for special relativity in the context of quantum mechanics. It was validated by accounting for the fine details of the hydrogen spectrum in a completely rigorous way.

The equation also implied the existence of a new form of matter, *antimatter*, previously unsuspected and unobserved and which was experimentally confirmed several years later. It also provided a *theoretical* justification for the introduction of several component wave functions in Pauli's phenomenological theory of spin. The wave functions in the Dirac theory are vectors of four complex numbers (known as bispinors), two of which resemble the Pauli wavefunction in the non-relativistic limit, in contrast to the Schrödinger equation which described wave functions of only one complex value. Moreover, in the limit of zero mass, the Dirac equation reduces to the Weyl equation.

Although Dirac did not at first fully appreciate the importance of his results, the entailed explanation of spin as a consequence of the union of quantum mechanics and relativity—and the eventual discovery of the positron—represents one of the great triumphs of theoretical physics. This accomplishment has been described as fully on a par with the works of Newton, Maxwell, and Einstein before him. In the context of quantum field theory, the Dirac equation is reinterpreted to describe quantum fields corresponding to spin $-\frac{1}{2}$ particles.

Mathematical Formulation

The Dirac equation in the form originally proposed by Dirac is:

$$\left(\beta mc^2 + c\left(\sum_{n=1}^{3}\alpha_n p_n\right)\right)\psi(x,t) = i\hbar\frac{\partial\psi(x,t)}{\partial t}$$

where $\psi=\psi(x,t)$ is the wave function for the electron of rest mass m with spacetime coordinates x,t. The p_1, p_2, p_3 are the components of the momentum, understood to be the momentum operator in the Schrödinger equation. Also, c is the speed of light, and $\hbar$ is the Planck constant divided by 2π. These fundamental physical constants reflect special relativity and quantum mechanics, respectively.

Dirac's purpose in casting this equation was to explain the behavior of the relativistically moving electron, and so to allow the atom to be treated in a manner consistent with relativity. His rather modest hope was that the corrections introduced this way might have a bearing on the problem of atomic spectra. Up until that time, attempts to make the old quantum theory of the atom compatible with the theory of relativity, attempts based on discretizing the angular momentum stored in the electron's possibly non-circular orbit of the atomic nucleus, had failed – and the new quantum mechanics of Heisenberg, Pauli, Jordan, Schrödinger, and Dirac himself had not developed sufficiently to treat this problem. Although Dirac's original intentions were satisfied, his equation had far deeper implications for the structure of matter and introduced new mathematical classes of objects that are now essential elements of fundamental physics.

The new elements in this equation are the 4 × 4 matrices α_k and β and the four-component wave function ψ. There are four components in ψ because the evaluation of it at any given point in configuration space is a bispinor. It is interpreted as a superposition of a spin-up electron, a spin-down electron, a spin-up positron, and a spin-down positron.

The 4 × 4 matrices α_k and β are all Hermitian and have squares equal to the identity matrix:

$$\alpha_i^2 = \beta^2 = I_4$$

and they all mutually anticommute (if i and j are distinct):

$$\alpha_i\alpha_j + \alpha_j\alpha_i = 0$$
$$\alpha_i\beta + \beta\alpha_i = 0$$

The single symbolic equation thus unravels into four coupled linear first-order partial differential equations for the four quantities that make up the wave function. These matrices and the form of the wave function have a deep mathematical significance. The algebraic structure represented by the gamma matrices had been created some 50 years earlier by the English mathematician W. K. Clifford. In turn, Clifford's ideas had emerged from the mid-19th-century work of the German mathematician Hermann Grassmann in his *Lineale Ausdehnungslehre* (*Theory of Linear Extensions*). The latter had been regarded as well-nigh incomprehensible by most of his contemporaries. The appearance of something so seemingly abstract, at such a late date, and in such a direct physical manner, is one of the most remarkable chapters in the history of physics.

Making the Schrödinger Equation Relativistic

The Dirac equation is superficially similar to the Schrödinger equation for a massive free particle:

$$-\frac{\hbar^2}{2m}\nabla^2\phi = i\hbar\frac{\partial}{\partial t}\phi.$$

The left side represents the square of the momentum operator divided by twice the mass, which is the non-relativistic kinetic energy. Because relativity treats space and time as a whole, a relativistic generalization of this equation requires that space and time derivatives must enter symmetrically as they do in the Maxwell equations that govern the behavior of light — the equations must be differentially of the *same order* in space and time. In relativity, the momentum and the energies are the space and time parts of a spacetime vector, the four-momentum, and they are related by the relativistically invariant relation,

$$E^2 = m^2c^4 + p^2c^2$$

which says that the length of this four-vector is proportional to the rest mass m. Substituting the operator equivalents of the energy and momentum from the Schrödinger theory, we get the Klein–Gordon equation describing the propagation of waves, constructed from relativistically invariant objects,

$$\left(-\frac{1}{c^2}\frac{\partial^2}{\partial t^2} + \nabla^2\right)\phi = \frac{m^2c^2}{\hbar^2}\phi$$

with the wave function ϕ being a relativistic scalar: a complex number which has the same numerical value in all frames of reference. Space and time derivatives both enter to second order. This has a telling consequence for the interpretation of the equation. Because the equation is second order in the time derivative, one must specify initial values both of the wave function itself and of its first time-derivative in order to solve definite problems. Since both may be specified more or less arbitrarily, the wave function cannot maintain its former role of determining the probability density of finding the electron in a given state of motion. In the Schrödinger theory, the probability density is given by the positive definite expression,

$$\rho = \phi^*\phi$$

and this density is convected according to the probability current vector,

$$J = -\frac{i\hbar}{2m}(\phi^*\nabla\phi - \phi\nabla\phi^*)$$

with the conservation of probability current and density following from the continuity equation:

$$\nabla\cdot J + \frac{\partial\rho}{\partial t} = 0.$$

The fact that the density is positive definite and convected according to this continuity equation implies that we may integrate the density over a certain domain and set the total to 1, and this condition will be maintained by the conservation law. A proper relativistic theory with a probability density current must also share this feature. Now, if we wish to maintain the notion of a convected density, then we must generalize the Schrödinger expression of the density and current so that space and time derivatives again enter symmetrically in relation to the scalar wave function. We are allowed to keep the Schrödinger expression for the current, but must replace the probability density by the symmetrically formed expression,

$$\rho = \frac{i\hbar}{2mc^2}(\psi^* \partial_t \psi - \psi \partial_t \psi^*).$$

which now becomes the 4th component of a spacetime vector, and the entire probability 4-current density has the relativistically covariant expression,

$$J^\mu = \frac{i\hbar}{2m}(\psi^* \partial^\mu \psi - \psi \partial^\mu \psi^*)$$

The continuity equation is as before. Everything is compatible with relativity now, but we see immediately that the expression for the density is no longer positive definite – the initial values of both ψ and $\partial_t \psi$ may be freely chosen, and the density may thus become negative, something that is impossible for a legitimate probability density. Thus, we cannot get a simple generalization of the Schrödinger equation under the naive assumption that the wave function is a relativistic scalar, and the equation it satisfies, second order in time.

Although it is not a successful relativistic generalization of the Schrödinger equation, this equation is resurrected in the context of quantum field theory, where it is known as the Klein–Gordon equation, and describes a spinless particle field (e.g. pi meson or Higgs boson). Historically, Schrödinger himself arrived at this equation before the one that bears his name but soon discarded it. In the context of quantum field theory, the indefinite density is understood to correspond to the *charge* density, which can be positive or negative, and not the probability density.

Dirac's Coup

Dirac thus thought to try an equation that was *first order* in both space and time. One could, for example, formally (i.e. by abuse of notation) take the relativistic expression for the energy

$$E = c\sqrt{p^2 + m^2c^2},$$

replace p by its operator equivalent, expand the square root in an infinite series of derivative operators, set up an eigenvalue problem, then solve the equation formally by iterations. Most physicists had little faith in such a process, even if it were technically possible.

As the story goes, Dirac was staring into the fireplace at Cambridge, pondering this problem, when he hit upon the idea of taking the square root of the wave operator thus:

$$\nabla^2 - \frac{1}{c^2}\frac{\partial^2}{\partial t^2} = \left(A\partial_x + B\partial_y + C\partial_z + \frac{i}{c}D\partial_t\right)\left(A\partial_x + B\partial_y + C\partial_z + \frac{i}{c}D\partial_t\right).$$

On multiplying out the right side we see that, in order to get all the cross-terms such as $\partial_x \partial_y$ to vanish, we must assume,

$$AB + BA = 0, \dots$$

with

$$A^2 = B^2 = \dots = 1.$$

Dirac, who had just then been intensely involved with working out the foundations of Heisenberg's matrix mechanics, immediately understood that these conditions could be met if *A*, *B*, *C* and *D* are *matrices*, with the implication that the wave function has *multiple components*. This immediately explained the appearance of two-component wave functions in Pauli's phenomenological theory of spin, something that up until then had been regarded as mysterious, even to Pauli himself. However, one needs at least 4 × 4 matrices to set up a system with the properties required — so the wave function had *four* components, not two, as in the Pauli theory, or one, as in the bare Schrödinger theory. The four-component wave function represents a new class of mathematical object in physical theories that makes its first appearance here.

Given the factorization in terms of these matrices, one can now write down immediately an equation,

$$\left(A\partial_x + B\partial_y + C\partial_z + \frac{i}{c} D\partial_t \right)\psi = \kappa\psi$$

with κ to be determined. Applying again the matrix operator on both sides yields,

$$\left(\nabla^2 - \frac{1}{c^2}\partial_t^2 \right)\psi = \kappa^2\psi.$$

On taking $\kappa = \frac{mc}{\hbar}$ we find that all the components of the wave function *individually* satisfy the relativistic energy–momentum relation. Thus the sought-for equation that is first-order in both space and time is,

$$\left(A\partial_x + B\partial_y + C\partial_z + \frac{i}{c} D\partial_t - \frac{mc}{\hbar} \right)\psi = 0.$$

Setting

$$A = i\beta\alpha_1, B = i\beta\alpha_2, C = i\beta\alpha_3, D = \beta,$$

and because $D^2 = \beta^2 = I_4$,

we get the Dirac equation as written above.

Covariant form and Relativistic Invariance

To demonstrate the relativistic invariance of the equation, it is advantageous to cast it into a form in which the space and time derivatives appear on an equal footing. New matrices are introduced as follows:

$$D = \gamma^0,$$
$$A = i\gamma^1, B = i\gamma^2, C = i\gamma^3,$$

and the equation takes the form (remembering the definition of the covariant components of the 4-gradient and especially that $\partial_0 = 1/c\partial_t$).

Dirac equation:

$$i\hbar\gamma^\mu\partial_\mu\psi - mc\psi = 0$$

where there is an implied summation over the values of the twice-repeated index μ = 0, 1, 2, 3, and ∂_μ is the 4-gradient. In practice one often writes the gamma matrices in terms of 2 × 2 sub-matrices taken from the Pauli matrices and the 2 × 2 identity matrix. Explicitly the standard representation is,

$$\gamma^0 = \begin{pmatrix} I_2 & 0 \\ 0 & -I_2 \end{pmatrix}, \gamma^1 = \begin{pmatrix} 0 & \sigma_x \\ -\sigma_x & 0 \end{pmatrix}, \gamma^2 = \begin{pmatrix} 0 & \sigma_y \\ -\sigma_y & 0 \end{pmatrix}, \gamma^3 = \begin{pmatrix} 0 & \sigma_z \\ -\sigma_z & 0 \end{pmatrix}.$$

The complete system is summarized using the Minkowski metric on spacetime in the form,

$$\{\gamma^\mu, \gamma^\nu\} = 2\eta^{\mu\nu} I_4$$

where the bracket expression,

$$\{a, b\} = ab + ba$$

denotes the anticommutator. These are the defining relations of a Clifford algebra over a pseudo-orthogonal 4-dimensional space with metric signature $(+ - - -)$ The specific Clifford algebra employed in the Dirac equation is known today as the Dirac algebra. Although not recognized as such by Dirac at the time the equation was formulated, in hindsight the introduction of this *geometric algebra* represents an enormous stride forward in the development of quantum theory.

The Dirac equation may now be interpreted as an eigenvalue equation, where the rest mass is proportional to an eigenvalue of the 4-momentum operator, the proportionality constant being the speed of light:

$$P_{\text{op}}\psi = mc\psi.$$

Using $\partial\!\!\!/ \stackrel{\text{def}}{=} \gamma^\mu\partial_\mu$ ($\partial\!\!\!/$ is pronounced "d-slash") , according to Feynman slash notation, the Dirac equation becomes:

$$i\hbar\partial\!\!\!/\psi - mc\psi = 0.$$

In practice, physicists often use units of measure such that $\hbar = c =$ 1,, known as natural units. The equation then takes the simple form.

Dirac Equation (Natural Units)

$$(i\partial - m)\psi = 0$$

A fundamental theorem states that if two distinct sets of matrices are given that both satisfy the Clifford relations, then they are connected to each other by a similarity transformation:

$$\gamma^{\mu'} = S^{-1}\gamma^{\mu}S.$$

If in addition the matrices are all unitary, as are the Dirac set, then *S* itself is unitary;

$$\gamma^{\mu'} = U^{+}\gamma^{\mu}U.$$

The transformation U is unique up to a multiplicative factor of absolute value 1. Let us now imagine a Lorentz transformation to have been performed on the space and time coordinates, and on the derivative operators, which form a covariant vector. For the operator $\gamma^{\mu}\partial_{\mu\mu}$ to remain invariant, the gammas must transform among themselves as a contravariant vector with respect to their spacetime index. These new gammas will themselves satisfy the Clifford relations, because of the orthogonality of the Lorentz transformation. By the fundamental theorem, we may replace the new set by the old set subject to a unitary transformation. In the new frame, remembering that the rest mass is a relativistic scalar, the Dirac equation will then take the form

$$(iU^{+}\gamma^{\mu}U\partial'_{\mu} - m)\psi(x',t') = 0$$
$$U^{+}(i\gamma^{\mu}\partial'_{\mu} - m)U\psi(x',t') = 0.$$

If we now define the transformed spinor

$$\psi' = U\psi$$

then we have the transformed Dirac equation in a way that demonstrates manifest relativistic invariance:

$$(i\gamma^{\mu}\partial'_{\mu} - m)\psi'(x',t') = 0$$

Thus, once we settle on any unitary representation of the gammas, it is final provided we transform the spinor according to the unitary transformation that corresponds to the given Lorentz transformation. The various representations of the Dirac matrices employed will bring into focus particular aspects of the physical content in the Dirac wave function. The representation shown here is known as the *standard* representation – in it, the wave function's upper two components go over into Pauli's 2-spinor wave function in the limit of low energies and small velocities in comparison to light.

The considerations above reveal the origin of the gammas in *geometry*, hearkening back to Grassmann's original motivation – they represent a fixed basis of unit vectors in spacetime. Similarly, products of the gammas such as $\gamma_{\mu}\gamma_{vv}$ represent *oriented surface elements*, and so on. With this in mind, we can find the form of the unit volume element on spacetime in terms of the gammas as follows. By definition, it is,

$$V = \frac{1}{4!}\epsilon_{\mu\nu\alpha\beta}\gamma^{\mu}\gamma^{\nu}\gamma^{\alpha}\gamma^{\beta}.$$

For this to be an invariant, the epsilon symbol must be a tensor, and so must contain a factor of $\sqrt{g}$, where g is the determinant of the metric tensor. Since this is negative, that factor is *imaginary*. Thus,

$$V = i\gamma^0\gamma^1\gamma^2\gamma^3.$$

This matrix is given the special symbol γ^5, owing to its importance when one is considering improper transformations of spacetime, that is, those that change the orientation of the basis vectors. In the standard representation, it is:

$$\gamma_5 = \begin{pmatrix} 0 & I_2 \\ I_2 & 0 \end{pmatrix}.$$

This matrix will also be found to anticommute with the other four Dirac matrices:

$$\gamma^5\gamma^\mu + \gamma^\mu\gamma^5 = 0$$

It takes a leading role when questions of *parity* arise because the volume element as a directed magnitude changes sign under a spacetime reflection. Taking the positive square root above thus amounts to choosing a handedness convention on spacetime.

Conservation of Probability Current

By defining the adjoint spinor,

$$\bar{\psi} = \psi^\dagger\gamma^0$$

where $\psi^\dagger$ is the conjugate transpose of ψ, and noticing that,

$$(\gamma^\mu)^\dagger\gamma^0 = \gamma^0\gamma^\mu,$$

we obtain, by taking the Hermitian conjugate of the Dirac equation and multiplying from the right by γ^0, the adjoint equation:

$$\bar{\psi}(-i\gamma^\mu\partial_\mu - m) = 0$$

where ∂_μ is understood to act to the left. Multiplying the Dirac equation by ψ from the left, and the adjoint equation by $\bar{\psi}$ from the right, and subtracting, produces the law of conservation of the Dirac current:

$$\partial_\mu\left(\bar{\psi}\gamma^\mu\psi\right) = 0.$$

Now we see the great advantage of the first-order equation over the one Schrödinger had tried – this is the conserved current density required by relativistic invariance, only now its 4th component is *positive definite* and thus suitable for the role of a probability density:

$$J^0 = \bar{\psi}\gamma^0\psi = \psi^\dagger\psi.$$

Because the probability density now appears as the fourth component of a relativistic vector and not a simple scalar as in the Schrödinger equation, it will be subject to the usual effects of the Lorentz transformations such as time dilation. Thus, for example, atomic processes that are observed as rates, will necessarily be adjusted in a way consistent with relativity, while those involving the measurement of energy and momentum, which themselves form a relativistic vector, will undergo parallel adjustment which preserves the relativistic covariance of the observed values.

Comparison with the Pauli Theory

The necessity of introducing half-integer spin goes back experimentally to the results of the Stern–Gerlach experiment. A beam of atoms is run through a strong inhomogeneous magnetic field, which then splits into N parts depending on the intrinsic angular momentum of the atoms. It was found that for silver atoms, the beam was split in two—the ground state therefore could not be integer, because even if the intrinsic angular momentum of the atoms were as small as possible, 1, the beam would be split into three parts, corresponding to atoms with $L_z = -1, 0, +1$. The conclusion is that silver atoms have net intrinsic angular momentum of $\frac{1}{2}$. Pauli set up a theory which explained this splitting by introducing a two-component wave function and a corresponding correction term in the Hamiltonian, representing a semi-classical coupling of this wave function to an applied magnetic field, as so in SI units: (Note that bold faced characters imply Euclidean vectors in 3 dimensions, whereas the Minkowski four-vector A_μ can be defined as $A_\mu = (\phi / c, -\mathbf{A})$.)

$$H = \frac{1}{2m}\left(\sigma \cdot (\mathbf{p} - e\mathbf{A})\right)^2 + e\phi.$$

Here $\mathbf{A}$ and ϕ represent the components of the electromagnetic four-potential in their standard SI units, and the three sigmas are the Pauli matrices. On squaring out the first term, a residual interaction with the magnetic field is found, along with the usual classical Hamiltonian of a charged particle interacting with an applied field in SI units:

$$H = \frac{1}{2m}(p - e\mathbf{A})^2 + e\phi - \frac{e\hbar}{2m}\sigma \cdot \mathbf{B}.$$

This Hamiltonian is now a 2 × 2 matrix, so the Schrödinger equation based on it must use a two-component wave function. On introducing the external electromagnetic 4-vector potential into the Dirac equation in a similar way, known as minimal coupling, it takes the form:

$$(\gamma^\mu (i\hbar\partial_\mu - eA_\mu) - mc)\psi = 0.$$

A second application of the Dirac operator will now reproduce the Pauli term exactly as before, because the spatial Dirac matrices multiplied by i, have the same squaring and commutation properties as the Pauli matrices. What is more, the value of the gyromagnetic ratio of the electron, standing in front of Pauli's new term, is explained from first principles. This was a major achievement of the Dirac equation and gave physicists great faith in its overall correctness. There is more however. The Pauli theory may be seen as the low energy limit of the Dirac theory in the following manner. First the equation is written in the form of coupled equations for 2-spinors with the SI units restored:

$$\begin{pmatrix} (mc^2 - E + e\phi) & c\sigma\cdot(\mathbf{p}-e\mathbf{A}) \\ -c\sigma\cdot(\mathbf{p}-e\mathbf{A}) & (mc^2+E-e\phi) \end{pmatrix}\begin{pmatrix} \psi_+ \\ \psi_- \end{pmatrix} = \begin{pmatrix} 0 \\ 0 \end{pmatrix}.$$

so

$$(E-e\phi)\psi_+ - c\sigma\cdot(\mathbf{p}-e\mathbf{A})\psi_- = mc^2\psi_+$$

$$-(E-e\phi)\psi_- + c\sigma\cdot(\mathbf{p}-e\mathbf{A})\psi_+ = mc^2\psi_-$$

Assuming the field is weak and the motion of the electron non-relativistic, we have the total energy of the electron approximately equal to its rest energy, and the momentum going over to the classical value,

$$E - e\phi \approx mc^2$$

$$\mathbf{p} \approx m\mathbf{v}$$

and so the second equation may be written:

$$\psi_- \approx \frac{1}{2mc}\sigma\cdot(\mathbf{p}-e\mathbf{A})\psi_+$$

which is of order $\frac{v}{c}$ – thus at typical energies and velocities, the bottom components of the Dirac spinor in the standard representation are much suppressed in comparison to the top components. Substituting this expression into the first equation gives after some rearrangement

$$(E-mc^2)\psi_+ = \frac{1}{2m}\left[\sigma\cdot(\mathbf{p}-e\mathbf{A})\right]^2\psi_+ + e\phi\psi_+ (E-mc^2)\psi_+ = \frac{1}{2m}\left[\sigma\cdot(\mathbf{p}-e\mathbf{A})\right]^2\psi_+ + e\phi\psi_+$$

The operator on the left represents the particle energy reduced by its rest energy, which is just the classical energy, so we recover Pauli's theory if we identify his 2-spinor with the top components of the Dirac spinor in the non-relativistic approximation. A further approximation gives the Schrödinger equation as the limit of the Pauli theory. Thus, the Schrödinger equation may be seen as the far non-relativistic approximation of the Dirac equation when one may neglect spin and work only at low energies and velocities. This also was a great triumph for the new equation, as it traced the mysterious *i* that appears in it, and the necessity of a complex wave function, back to the geometry of spacetime through the Dirac algebra. It also highlights why the Schrödinger equation, although superficially in the form of a diffusion equation, actually represents the propagation of waves.

It should be strongly emphasized that this separation of the Dirac spinor into large and small components depends explicitly on a low-energy approximation. The entire Dirac spinor represents an *irreducible* whole, and the components we have just neglected to arrive at the Pauli theory will bring in new phenomena in the relativistic regime – antimatter and the idea of creation and annihilation of particles.

Comparison with the Weyl Theory

In the limit $m \rightarrow 0$, the Dirac equation reduces to the Weyl equation, which describes relativistic massless spin-½ particles.

Dirac Lagrangian

Both the Dirac equation and the Adjoint Dirac equation can be obtained from (varying) the action with a specific Lagrangian density that is given by:

$$\mathcal{L} = i\hbar c\bar{\psi}\gamma^{\mu}\partial_{\mu}\psi - mc^2\bar{\psi}\psi$$

If one varies this with respect to ψ one gets the Adjoint Dirac equation. Meanwhile, if one varies this with respect to ψ one gets the Dirac equation.

Physical Interpretation

Identification of Observables

The critical physical question in a quantum theory is—what are the physically observable quantities defined by the theory? According to the postulates of quantum mechanics, such quantities are defined by Hermitian operators that act on the Hilbert space of possible states of a system. The eigenvalues of these operators are then the possible results of measuring the corresponding physical quantity. In the Schrödinger theory, the simplest such object is the overall Hamiltonian, which represents the total energy of the system. If we wish to maintain this interpretation on passing to the Dirac theory, we must take the Hamiltonian to be,

$$H = \gamma^0\left[mc^2 + c\gamma^k\left(p_k - qA_k\right)\right] + qA^0.$$

where, as always, there is an implied summation over the twice-repeated index $k = 1, 2, 3$. This looks promising, because we see by inspection the rest energy of the particle and, in case $A = 0$, the energy of a charge placed in an electric potential qA^0. What about the term involving the vector potential? In classical electrodynamics, the energy of a charge moving in an applied potential is

$$H = c\sqrt{\left(p - qA\right)^2 + m^2c^2} + qA^0.$$

Thus, the Dirac Hamiltonian is fundamentally distinguished from its classical counterpart, and we must take great care to correctly identify what is observable in this theory. Much of the apparently paradoxical behavior implied by the Dirac equation amounts to a misidentification of these observables.

Hole Theory

The negative E solutions to the equation are problematic, for it was assumed that the particle has a positive energy. Mathematically speaking, however, there seems to be no reason for us to reject the negative-energy solutions. Since they exist, we cannot simply ignore them, for once we include the interaction between the electron and the electromagnetic field, any electron placed in a positive-energy eigenstate would decay into negative-energy eigenstates of successively lower energy. Real electrons obviously do not behave in this way, or they would disappear by emitting energy in the form of photons.

To cope with this problem, Dirac introduced the hypothesis, known as hole theory, that the vacuum is the many-body quantum state in which all the negative-energy electron eigenstates are occupied. This description of the vacuum as a "sea" of electrons is called the Dirac sea. Since the Pauli exclusion principle forbids electrons from occupying the same state, any additional electron would be forced to occupy a positive-energy eigenstate, and positive-energy electrons would be forbidden from decaying into negative-energy eigenstates.

If an electron is forbidden from simultaneously occupying positive-energy and negative-energy eigenstates, then the feature known as Zitterbewegung, which arises from the interference of positive-energy and negative-energy states, would have to be considered to be an unphysical prediction of time-dependent Dirac theory. This conclusion may be inferred from the explanation of hole theory given in the preceding paragraph. Recent results have been published in Nature in which the Zitterbewegung feature was simulated in a trapped-ion experiment. This experiment impacts the hole interpretation if one infers that the physics-laboratory experiment is not merely a check on the mathematical correctness of a Dirac-equation solution but the measurement of a real effect whose detectability in electron physics is still beyond reach.

Dirac further reasoned that if the negative-energy eigenstates are incompletely filled, each unoccupied eigenstate – called a hole – would behave like a positively charged particle. The hole possesses a *positive* energy since energy is required to create a particle–hole pair from the vacuum. As noted above, Dirac initially thought that the hole might be the proton, but Hermann Weyl pointed out that the hole should behave as if it had the same mass as an electron, whereas the proton is over 1800 times heavier. The hole was eventually identified as the positron, experimentally discovered by Carl Anderson in 1932.

It is not entirely satisfactory to describe the "vacuum" using an infinite sea of negative-energy electrons. The infinitely negative contributions from the sea of negative-energy electrons have to be canceled by an infinite positive "bare" energy and the contribution to the charge density and current coming from the sea of negative-energy electrons is exactly canceled by an infinite positive "jellium" background so that the net electric charge density of the vacuum is zero. In quantum field theory, a Bogoliubov transformation on the creation and annihilation operators (turning an occupied negative-energy electron state into an unoccupied positive energy positron state and an unoccupied negative-energy electron state into an occupied positive energy positron state) allows us to bypass the Dirac sea formalism even though, formally, it is equivalent to it.

In certain applications of condensed matter physics, however, the underlying concepts of "hole theory" are valid. The sea of conduction electrons in an electrical conductor, called a Fermi sea, contains electrons with energies up to the chemical potential of the system. An unfilled state in the Fermi sea behaves like a positively charged electron, though it is referred to as a "hole" rather than a "positron". The negative charge of the Fermi sea is balanced by the positively charged ionic lattice of the material.

Quantum Field Theory

In quantum field theories such as quantum electrodynamics, the Dirac field is subject to a process of second quantization, which resolves some of the paradoxical features of the equation.

Other Formulations

The Dirac equation can be formulated in a number of other ways.

Differential Equation in One Real Component

Generically (if a certain linear function of electromagnetic field does not vanish identically), three out of four components of the spinor function in the Dirac equation can be algebraically eliminated, yielding an equivalent fourth-order partial differential equation for just one component. Furthermore, this remaining component can be made real by a gauge transform.

Curved Spacetime

This topic has developed the Dirac equation in flat spacetime according to special relativity. It is possible to formulate the Dirac equation in curved spacetime.

Algebra of Physical Space

This topic developed the Dirac equation using four vectors and Schrödinger operators. The Dirac equation in the algebra of physical space uses a Clifford algebra over the real numbers, a type of geometric algebra.

Polar Form

For the Dirac spinor, it is possible to show that one can always find local Lorentz transformations for which the spinor is written in the so-called polar form, that is the form manifesting only two physical degrees of freedom, given by the scalar and pseudo-scalar bi-linear quantities: the Dirac spinor field equation can always be written in the corresponding polar form, which is the form specifying all derivatives of the scalar and pseudo-scalar bi-linear quantities themselves. In this formulation, the Dirac spinor field equation (which are four complex equations, and so eight equations in total) is converted into an equivalent system of two real vector equations (which are two 4-dimensional equations, and so again eight equations in total). This is what is known as the polar form of the Dirac equation.

Solutions to the Dirac Equation

As with the Schrödinger equation, the simplest solutions of the Dirac equation are those for a free particle. They are also quite important to understand. We will find that each component of the Dirac spinor represents a state of a free particle at rest that we can interpret fairly easily.

We can show that a free particle solution can be written as a constant spinor times the usual free particle exponential. Start from the Dirac equation and attempt to develop an equation to show that each component has the free particle exponential. We will do this by making a second order differential equation, which turns out to be the Klein-Gordon equation.

$$\left(\gamma_\mu \frac{\partial}{\partial x_\mu} + \frac{m}{\hbar}\right)\psi = 0$$

$$\left(\gamma_\upsilon \frac{\partial}{\partial x_\upsilon}\left(\gamma_\mu \frac{\partial}{\partial x_\mu} + \frac{mc}{\hbar}\right)\right)\psi = 0$$

$$\gamma_\upsilon \frac{\partial}{\partial x_\upsilon}\left(\gamma_\mu \frac{\partial}{\partial x_\mu}\psi + \gamma_\upsilon \frac{\partial}{\partial x_\mu}\frac{mc}{\hbar}\psi = 0\right.$$

$$\gamma_\upsilon\gamma_\mu \frac{\partial}{\partial x_\upsilon}\frac{\partial}{\partial x_\mu}\psi + \frac{mc}{\hbar}\gamma_\upsilon \frac{\partial}{\partial x_\upsilon}\psi = 0$$

$$\gamma_\upsilon\gamma_\mu \frac{\partial}{\partial x_\upsilon}\frac{\partial}{\partial x_\mu}\psi - \left(\frac{mc}{\hbar}\right)^2 \psi = 0$$

$$\gamma_\mu\gamma_\upsilon \frac{\partial}{\partial x_\mu}\frac{\partial}{\partial x_\upsilon}\psi - \left(\frac{mc}{\hbar}\right)^2 \psi = 0$$

$$(\gamma_\upsilon\gamma_\mu + \gamma_\mu\gamma_\upsilon)\frac{\partial}{\partial x_\mu}\frac{\partial}{\partial x_\upsilon}\psi - 2\left(\frac{mc}{\hbar}\right)^2 \psi = 0$$

$$2\delta_{\upsilon\mu}\frac{\partial}{\partial x_\mu}\frac{\partial}{\partial x_\upsilon}\psi - 2\left(\frac{mc}{\hbar}\right)^2 \psi = 0$$

$$2\frac{\partial}{\partial x_\mu}\frac{\partial}{\partial x_\mu}\psi - 2\left(\frac{mc}{\hbar}\right)^2 \psi = 0$$

The free electron solutions all satisfy the wave equation.

$$\left(\Box - \left(\frac{mc}{\hbar}\right)^2\right)\psi = 0$$

Because we have eliminated the γ matrices from the equation, this is an equation for each component of the Dirac spinor ψ. Each component satisfies the wave (Klein-Gordon) equation and a solution can be written as a constant spinor times the usual exponential representing a wave,

$$\psi_{\vec{p}} = u_{\vec{p}} e^{i(\vec{p}\cdot\vec{x} - Et)/\hbar}$$

Plugging this into the equation, we get a relation between the momentum and the energy.

$$\frac{-p^2}{\hbar}+\frac{E^2}{\hbar^2 C^2}-\left(\frac{mc}{\hbar}\right)^2=0$$

$$-p^2c^2+E^2-m^2c^4=0$$
$$E^2=p^2c^2+m^2c^4$$
$$E=\pm\sqrt{p^2c^2+m^2c^4}$$

Note that the momentum operator is clearly still $\frac{\hbar}{i}\vec{\nabla}$ and the energy operator is still,

$$i\hbar\frac{\partial}{\partial t}$$

There is no coupling between the different components in this equation, but, we will see that (unlike the equation differentiated again) the Dirac equation will give us relations between the components of the constant spinor. Again, the solution can be written as a constant spinor, which may depend on momentum $u_{\vec{p}}$ times the exponential.

$$\psi_{\vec{p}}(x)=u_{\vec{p}}e^{i(\vec{p}\cdot\vec{x}-Et)/\hbar}$$

We should normalize the state if we want to describe one particle per unit volume: $\psi^{\dagger}\psi=\frac{1}{V}$ We haven't learned much about what each component represents yet. We also have the plus or minus in the relation $E=\pm\sqrt{p^2c^2+m^2c^4}$ to deal with. The solutions for a free particle at rest will tell us more about what the different components mean.

The Feynman Diagrams

In theoretical physics, Feynman diagrams are pictorial representations of the mathematical expressions describing the behavior of subatomic particles. The scheme is named after its inventor, American physicist Richard Feynman, and was first introduced in 1948. The interaction of sub-atomic particles can be complex and difficult to understand intuitively. Feynman diagrams give a simple visualization of what would otherwise be an arcane and abstract formula. As David Kaiser writes, "since the middle of the 20th century, theoretical physicists have increasingly turned to this tool to help them undertake critical calculations", and so "Feynman diagrams have revolutionized nearly every aspect of theoretical physics". While the diagrams are applied primarily to quantum field theory, they can also be used in other fields, such as solid-state theory.

Feynman used Ernst Stueckelberg's interpretation of the positron as if it were an electron moving backward in time. Thus, antiparticles are represented as moving backward along the time axis in Feynman diagrams.

The calculation of probability amplitudes in theoretical particle physics requires the use of rather large and complicated integrals over a large number of variables. These integrals do, however, have a regular structure, and may be represented graphically as Feynman diagrams.

A Feynman diagram is a contribution of a particular class of particle paths, which join and split as described by the diagram. More precisely, and technically, a Feynman diagram is a graphical representation of a perturbative contribution to the transition amplitude or correlation function of a quantum mechanical or statistical field theory. Within the canonical formulation of quantum field theory, a Feynman diagram represents a term in the Wick's expansion of the perturbative *S*-matrix. Alternatively, the path integral formulation of quantum field theory represents the transition amplitude as a weighted sum of all possible histories of the system from the initial to the final state, in terms of either particles or fields. The transition amplitude is then given as the matrix element of the *S*-matrix between the initial and the final states of the quantum system.

A Feynman diagram represents a perturbative contribution to the amplitude of a quantum transition from some initial quantum state to some final quantum state.

For example, in the process of electron-positron annihilation the initial state is one electron and one positron, the final state: two photons.

The initial state is often assumed to be at the left of the diagram and the final state at the right (although other conventions are also used quite often).

A Feynman diagram consists of points, called vertices, and lines attached to the vertices.

The particles in the initial state are depicted by lines sticking out in the direction of the initial state (e.g., to the left), the particles in the final state are represented by lines sticking out in the direction of the final state (e.g., to the right).

In QED there are two types of particles: matter particles such as electrons or positrons (called fermions) and exchange particles (called gauge bosons). They are represented in Feynman diagrams as follows:

1. Electron in the initial state is represented by a solid line, with an arrow indicating the spin of the particle e.g. pointing toward the vertex (→•).
2. Electron in the final state is represented by a line, with an arrow indicating the spin of the particle e.g. pointing away from the vertex: (•→).
3. Positron in the initial state is represented by a solid line, with an arrow indicating the spin of the particle e.g. pointing away from the vertex: (←•).
4. Positron in the final state is represented by a line, with an arrow indicating the spin of the particle e.g. pointing toward the vertex: (•←).
5. Virtual Photon in the initial and the final state is represented by a wavy line (~• and •~).

In QED a vertex always has three lines attached to it: one bosonic line, one fermionic line with arrow toward the vertex, and one fermionic line with arrow away from the vertex.

The vertices might be connected by a bosonic or fermionic propagator. A bosonic propagator is represented by a wavy line connecting two vertices (•~•). A fermionic propagator is represented by a solid line (with an arrow in one or another direction) connecting two vertices, (•←•).

The number of vertices gives the order of the term in the perturbation series expansion of the transition amplitude.

Electron–positron Annihilation Example

The electron–positron annihilation interaction:

$$e^{+}e^{-} \rightarrow 2\gamma$$

has a contribution from the second order Feynman diagram shown adjacent:

In the initial state (at the bottom; early time) there is one electron (e^{-}) and one positron e^{+} and in the final state (at the top; late time) there are two photons (γ).

Representation of Physical Reality

In their presentations of fundamental interactions, written from the particle physics perspective, Gerard ‘t Hooft and Martinus Veltman gave good arguments for taking the original, non-regularized Feynman diagrams as the most succinct representation of our present knowledge about the physics of quantum scattering of fundamental particles. Their motivations are consistent with the convictions of James Daniel Bjorken and Sidney Drell:

The Feynman graphs and rules of calculation summarize quantum field theory in a form in close contact with the experimental numbers one wants to understand. Although the statement of the theory in terms of graphs may imply perturbation theory, use of graphical methods in the many-body problem shows that this formalism is flexible enough to deal with phenomena of nonperturbative characters. Some modification of the Feynman rules of calculation may well outlive the elaborate mathematical structure of local canonical quantum field theory.

So far there are no opposing opinions. In quantum field theories the Feynman diagrams are obtained from Lagrangian by Feynman rules.

Dimensional regularization is a method for regularizing integrals in the evaluation of Feynman diagrams; it assigns values to them that are meromorphic functions of an auxiliary complex parameter *d*, called the dimension. Dimensional regularization writes a Feynman integral as an integral depending on the spacetime dimension *d* and spacetime points.

Particle-path Interpretation

A Feynman diagram is a representation of quantum field theory processes in terms of particle interactions. The particles are represented by the lines of the diagram, which can be squiggly or straight, with an arrow or without, depending on the type of particle. A point where lines connect to other lines is a *vertex*, and this is where the particles meet and interact: by emitting or absorbing new particles, deflecting one another, or changing type.

There are three different types of lines: *internal lines* connect two vertices, *incoming lines* extend from "the past" to a vertex and represent an initial state, and *outgoing lines* extend from a vertex to "the future" and represent the final state (the latter two are also known as *external lines*). Traditionally, the bottom of the diagram is the past and the top the future; other times, the past is to the left and the future to the right. When calculating correlation functions instead of scattering amplitudes, there is no past and future and all the lines are internal. The particles then begin and end on little x's, which represent the positions of the operators whose correlation is being calculated.

Feynman diagrams are a pictorial representation of a contribution to the total amplitude for a process that can happen in several different ways. When a group of incoming particles are to scatter off each other, the process can be thought of as one where the particles travel over all possible paths, including paths that go backward in time.

Feynman diagrams are often confused with spacetime diagrams and bubble chamber images because they all describe particle scattering. Feynman diagrams are graphs that represent the interaction of particles rather than the physical position of the particle during a scattering process. Unlike a bubble chamber picture, only the sum of all the Feynman diagrams represent any given particle interaction; particles do not choose a particular diagram each time they interact. The law of summation is in accord with the principle of superposition—every diagram contributes to the total amplitude for the process.

Canonical Quantization Formulation

The probability amplitude for a transition of a quantum system from the initial state $|i\rangle$ to the final state $|f\rangle$ is given by the matrix element:

$$S_{fi} = \langle f | S | i \rangle ,$$

where S is the S-matrix.

In the canonical quantum field theory the S-matrix is represented within the interaction picture by the perturbation series in the powers of the interaction Lagrangian,

$$S = \sum_{n=0}^{\infty} \frac{i^n}{n!} \int \prod_{j=1}^{n} d^4 x_j T \prod_{j=1}^{n} L_v(x_j) \equiv \sum_{n=0}^{\infty} S^{(n)} ,$$

where L_v is the interaction Lagrangian and T signifies the time-ordered product of operators.

A Feynman diagram is a graphical representation of a term in the Wick's expansion of the time-ordered product in the nth order term $S^{(n)}$ of the S-matrix,

$$T \prod_{j=1}^{n} L_v(x_j) = \sum_{\substack{\text{all possible} \\ \text{contractions}}} (\pm) N \prod_{j=1}^{n} L_v(x_j) ,$$

where N signifies the normal-product of the operators and $(\pm)$ takes care of the possible sign change when commuting the fermionic operators to bring them together for a contraction (a propagator).

Feynman Rules

The diagrams are drawn according to the Feynman rules, which depend upon the interaction Lagrangian. For the QED interaction Lagrangian,

$$L_v = -g\bar{\psi}\gamma^{\mu}\psi A_{\mu}$$

describing the interaction of a fermionic field ψ with a bosonic gauge field A_{μ}, the Feynman rules can be formulated in coordinate space as follows:

1. Each integration coordinate x_j is represented by a point (sometimes called a vertex);
2. A bosonic propagator is represented by a wiggly line connecting two points;
3. A fermionic propagator is represented by a solid line connecting two points;
4. A bosonic field $A_{\mu}(x_i)$ is represented by a wiggly line attached to the point x_i;
5. A fermionic field $\psi(x_i)$ is represented by a solid line attached to the point x_i with an arrow toward the point;
6. An anti-fermionic field $\bar{\psi}(x_i)$ is represented by a solid line attached to the point x_i with an arrow away from the point;

Example of Second Order Processes in QED

The second order perturbation term in the S-matrix is:

$$S^{(2)} = \frac{(ie)^2}{2!}\int d^4x d^4x' T\bar{\psi}(x)\gamma^{\mu}\psi(x)A_{\mu}(x)\bar{\psi}(x')\gamma^{\nu}\psi(x')A_{\nu}(x').$$

Scattering of Fermions

The Feynman diagram of the term.

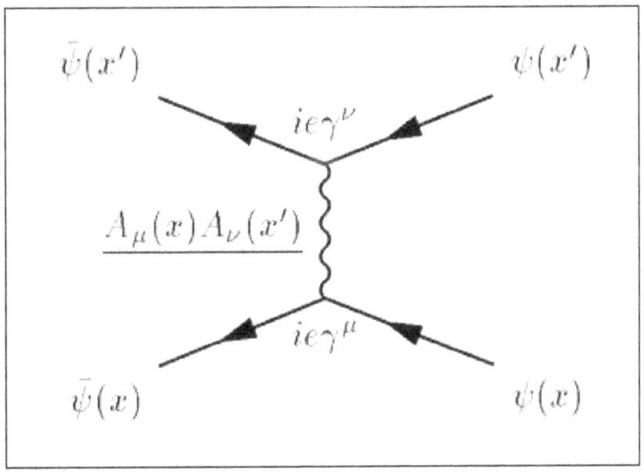

$$N\bar{\psi}(x)\gamma^{\mu}\psi(x)\bar{\psi}(x')\gamma^{\nu}\psi(x')\underline{A_{\mu}(x)A_{\nu}(x')},$$

The Wick's expansion of the integrand gives (among others) the following term.

Where:

$$\underline{A_\mu(x)A_\nu(x')} = \int \frac{d^4k}{(2\pi)^4} \frac{-ig_{\mu\nu}}{k^2 + io} e^{-ik(x-x')}$$

is the electromagnetic contraction (propagator) in the Feynman gauge. This term is represented by the Feynman diagram at the right. This diagram gives contributions to the following processes:

1. e^- e^- scattering (initial state at the right, final state at the left of the diagram);
2. e^+ e^+ scattering (initial state at the left, final state at the right of the diagram);
3. e^- e^+ scattering (initial state at the bottom/top, final state at the top/bottom of the diagram).

Compton Scattering and Annihilation/Generation of e^- e^+ Pairs

Another interesting term in the expansion is,

$$N\bar{\psi}(x)\gamma^\mu \underline{\psi(x)\bar{\psi}(x')}\gamma^\nu \psi(x')A_\mu(x)A_\nu(x'),$$

where,

$$\underline{\psi(x)\bar{\psi}(x')} = \int \frac{d^4p}{(2\pi)^4} \frac{i}{\gamma p - m + io} e^{-ip(x-x')}$$

is the fermionic contraction (propagator).

Path Integral Formulation

In a path integral, the field Lagrangian, integrated over all possible field histories, defines the probability amplitude to go from one field configuration to another. In order to make sense, the field theory should have a well-defined ground state, and the integral should be performed a little bit rotated into imaginary time, i.e. a Wick rotation.

Scalar Field Lagrangian

A simple example is the free relativistic scalar field in d dimensions, whose action integral is:

$$S = \int \tfrac{1}{2}\partial_\mu\phi\partial^\mu\phi d^d x.$$

The probability amplitude for a process is:

$$\int_A^B e^{iS}\, D\phi,$$

where A and B are space-like hypersurfaces that define the boundary conditions. The collection of all the $\varphi(A)$ on the starting hypersurface give the initial value of the field, analogous to the starting position for a point particle, and the field values $\varphi(B)$ at each point of the final hypersurface defines the final field value, which is allowed to vary, giving a different amplitude to end up at different values. This is the field-to-field transition amplitude.

The path integral gives the expectation value of operators between the initial and final state:

$$\int_A^B e^{iS}\phi(x_1)\cdots\phi(x_n)D\phi = \langle A|\phi(x_1)\cdots\phi(x_n)|B\rangle,$$

and in the limit that A and B recede to the infinite past and the infinite future, the only contribution that matters is from the ground state (this is only rigorously true if the path-integral is defined slightly rotated into imaginary time). The path integral can be thought of as analogous to a probability distribution, and it is convenient to define it so that multiplying by a constant doesn't change anything:

$$\frac{\int e^{iS}\phi(x_1)\cdots\phi(x_n)D\phi}{\int e^{iS}\,D\phi} = \langle 0|\phi(x_1)\cdots\phi(x_n)|0\rangle.$$

The normalization factor on the bottom is called the *partition function* for the field, and it coincides with the statistical mechanical partition function at zero temperature when rotated into imaginary time.

The initial-to-final amplitudes are ill-defined if one thinks of the continuum limit right from the beginning, because the fluctuations in the field can become unbounded. So the path-integral can be thought of as on a discrete square lattice, with lattice spacing a and the limit $a \to 0$ should be taken carefully. If the final results do not depend on the shape of the lattice or the value of a, then the continuum limit exists.

On a Lattice

On a lattice, (i), the field can be expanded in Fourier modes:

$$\phi(x) = \int \frac{dk}{(2\pi)^d}\phi(k)e^{ik\cdot x} = \int_k \phi(k)e^{ikx}$$

Here the integration domain is over k restricted to a cube of side length $\frac{2\pi}{a}$ so that large values of k are not allowed. It is important to note that the k-measure contains the factors of 2π from Fourier transforms, this is the best standard convention for k-integrals in QFT. The lattice means that fluctuations at large k are not allowed to contribute right away, they only start to contribute in the limit $a \to 0$. Sometimes, instead of a lattice, the field modes are just cut off at high values of k instead.

It is also convenient from time to time to consider the space-time volume to be finite, so that the k modes are also a lattice. This is not strictly as necessary as the space-lattice limit, because interactions in k are not localized, but it is convenient for keeping track of the factors in front of the k-integrals and the momentum-conserving delta functions that will arise.

On a lattice, (ii), the action needs to be discretized:

$$S = \sum_{\langle x,y\rangle} \tfrac{1}{2}\left(\phi(x) - \phi(y)\right)^2,$$

where $\langle x,y\rangle$ is a pair of nearest lattice neighbors x and y. The discretization should be thought of as defining what the derivative $\partial_\mu \varphi$ means.

In terms of the lattice Fourier modes, the action can be written:

$$S = \int_k \left(\left(1 - \cos(k_1)\right) + \left(1 - \cos(k_2)\right) + \cdots + \left(1 - \cos(k_d)\right)\right)\phi_k^* \phi^k .$$

For k near zero this is:

$$S = \int_k \tfrac{1}{2}k^2 \left|\phi(k)\right|^2 .$$

Now we have the continuum Fourier transform of the original action. In finite volume, the quantity $d^d k$ is not infinitesimal, but becomes the volume of a box made by neighboring Fourier modes, or $(\frac{2\pi}{V})^d$.

The field φ is real-valued, so the Fourier transform obeys:

$$\phi(k)^* = \phi(-k).$$

In terms of real and imaginary parts, the real part of $\varphi(k)$ is an even function of k, while the imaginary part is odd. The Fourier transform avoids double-counting, so that it can be written:

$$S = \int_k \tfrac{1}{2}k^2 \phi(k)\phi(-k)$$

over an integration domain that integrates over each pair $(k, -k)$ exactly once.

For a complex scalar field with action

$$S = \int \tfrac{1}{2}\partial_\mu \phi^* \partial^\mu \phi d^d x$$

the Fourier transform is unconstrained:

$$S = \int_k \tfrac{1}{2}k^2 \left|\phi(k)\right|^2$$

and the integral is over all k.

Integrating over all different values of $\varphi(x)$ is equivalent to integrating over all Fourier modes, because taking a Fourier transform is a unitary linear transformation of field coordinates. When you change coordinates in a multidimensional integral by a linear transformation, the value of the new integral is given by the determinant of the transformation matrix. If

$$y_i = A_{ij} x_j,$$

then

$$\det(A)\int dx_1 dx_2 \cdots dx_n = \int dy_1 dy_2 \cdots dy_n .$$

If A is a rotation, then

$$A^{\mathrm{T}} A = I$$

so that det $A = \pm 1$, , and the sign depends on whether the rotation includes a reflection or not.

The matrix that changes coordinates from $\varphi(x)$ to $\varphi(k)$ can be read off from the definition of a Fourier transform.

$$A_{kx} = e^{ikx}$$

and the Fourier inversion theorem tells you the inverse:

$$A^{-1}_{kx} = e^{-ikx}$$

which is the complex conjugate-transpose, up to factors of 2π. On a finite volume lattice, the determinant is nonzero and independent of the field values.

$$\det A = 1$$

and the path integral is a separate factor at each value of k.

$$\int \exp\left(\frac{i}{2}\sum_k k^2 \phi^*(k)\phi(k)\right) D\phi = \prod_k \int_{\phi_k} e^{\frac{i}{2}k^2 |\phi_k|^2 d^d k}$$

The factor $d^d k$ is the infinitesimal volume of a discrete cell in k-space, in a square lattice box

$$d^d k = \left(\frac{1}{L}\right)^d,$$

where L is the side-length of the box. Each separate factor is an oscillatory Gaussian, and the width of the Gaussian diverges as the volume goes to infinity.

In imaginary time, the *Euclidean action* becomes positive definite, and can be interpreted as a probability distribution. The probability of a field having values φ_k is,

$$e^{\int_k -\frac{1}{2}k^2 \phi_k^* \phi_k} = \prod_k e^{-k^2 |\phi_k|^2 d^d k}.$$

The expectation value of the field is the statistical expectation value of the field when chosen according to the probability distribution:

$$\langle \phi(x_1)\cdots\phi(x_n)\rangle = \frac{\int e^{-S}\phi(x_1)\cdots\phi(x_n)D\phi}{\int e^{-S}\,D\phi}$$

Since the probability of φ_k is a product, the value of $\varphi_k$at each separate value of k is independently Gaussian distributed. The variance of the Gaussian is $\frac{1}{k^2 d^d k}$ which is formally infinite, but that just means that the fluctuations are unbounded in infinite volume. In any finite volume, the integral is replaced by a discrete sum, and the variance of the integral is $\frac{V}{k^2}$.

Monte Carlo

The path integral defines a probabilistic algorithm to generate a Euclidean scalar field configuration. Randomly pick the real and imaginary parts of each Fourier mode at wavenumber k to be a Gaussian random variable with variance $\frac{1}{k^2}$. This generates a configuration $\varphi_C(k)$at random, and the Fourier transform gives $\varphi_C(x)$For real scalar fields, the algorithm must generate only one of each pair $\varphi(k)$ $\varphi(-k)$ and make the second the complex conjugate of the first.

To find any correlation function, generate a field again and again by this procedure, and find the statistical average:

$$\langle \phi(x_1)\cdots\phi(x_n)\rangle = \lim_{|C|\to\infty} \frac{\sum_C \phi_C(x_1)\cdots\phi_C(x_n)}{|C|}$$

where $|C|$ is the number of configurations, and the sum is of the product of the field values on each configuration. The Euclidean correlation function is just the same as the correlation function in statistics or statistical mechanics. The quantum mechanical correlation functions are an analytic continuation of the Euclidean correlation functions.

For free fields with a quadratic action, the probability distribution is a high-dimensional Gaussian, and the statistical average is given by an explicit formula. But the Monte Carlo method also works well for bosonic interacting field theories where there is no closed form for the correlation functions.

Scalar Propagator

Each mode is independently Gaussian distributed. The expectation of field modes is easy to calculate:

$$\langle \phi_k \phi_{k'} \rangle = 0$$

for $k \neq k'$ since then the two Gaussian random variables are independent and both have zero mean.

$$\langle \phi_k \phi_k \rangle = \frac{V}{k^2}$$

in finite volume V, when the two k-values coincide, since this is the variance of the Gaussian. In the infinite volume limit,

$$\langle \phi(k)\phi(k')\rangle = \delta(k-k')\frac{1}{k^2}$$

Strictly speaking, this is an approximation: the lattice propagator is:

$$\langle \phi(k)\phi(k')\rangle = \delta(k-k')\frac{1}{2\left(d-\cos(k_1)+\cos(k_2)\cdots+\cos(k_d)\right)}$$

But near k = 0, for field fluctuations long compared to the lattice spacing, the two forms coincide.

It is important to emphasize that the delta functions contain factors of 2π, so that they cancel out the 2π factors in the measure for k integrals.

$$\delta(k) = (2\pi)^d \delta_D(k_1)\delta_D(k_2)\cdots\delta_D(k_d)$$

where $\delta_D(k)$ is the ordinary one-dimensional Dirac delta function. This convention for delta-functions is not universal—some authors keep the factors of 2π in the delta functions (and in the k-integration) explicit.

Equation of Motion

The form of the propagator can be more easily found by using the equation of motion for the field. From the Lagrangian, the equation of motion is:

$$\partial_\mu \partial^\mu \phi = 0$$

and in an expectation value, this says:

$$\partial_\mu \partial^\mu \langle \phi(x)\phi(y)\rangle = 0$$

Where the derivatives act on x, and the identity is true everywhere except when x and y coincide, and the operator order matters. The form of the singularity can be understood from the canonical commutation relations to be a delta-function. Defining the (Euclidean) *Feynman propagator* Δ as the Fourier transform of the time-ordered two-point function (the one that comes from the path-integral):

$$\partial^2 \Delta(x) = i\delta(x)$$

So that:

$$\Delta(k) = \frac{i}{k^2}$$

If the equations of motion are linear, the propagator will always be the reciprocal of the quadratic-form matrix that defines the free Lagrangian, since this gives the equations of motion. The factor of i disappears in the Euclidean theory.

Wick Theorem

Because each field mode is an independent Gaussian, the expectation values for the product of many field modes obeys *Wick's theorem*:

$$\langle \phi(k_1)\phi(k_2)\cdots\phi(k_n)\rangle$$

is zero unless the field modes coincide in pairs. This means that it is zero for an odd number of φ, and for an even number of φ, it is equal to a contribution from each pair separately, with a delta function.

$$\langle \phi(k_1)\cdots\phi(k_{2n})\rangle = \sum\prod_{i,j}\frac{\delta(k_i - k_j)}{k_i^2}$$

where the sum is over each partition of the field modes into pairs, and the product is over the pairs. For example,

$$\langle \phi(k_1)\phi(k_2)\phi(k_3)\phi(k_4)\rangle = \frac{\delta(k_1-k_2)}{k_1^2}\frac{\delta(k_3-k_4)}{k_3^2}+\frac{\delta(k_1-k_3)}{k_3^2}\frac{\delta(k_2-k_4)}{k_2^2}+\frac{\delta(k_1-k_4)}{k_1^2}\frac{\delta(k_2-k_3)}{k_2^2}$$

An interpretation of Wick's theorem is that each field insertion can be thought of as a dangling line, and the expectation value is calculated by linking up the lines in pairs, putting a delta function factor that ensures that the momentum of each partner in the pair is equal, and dividing by the propagator.

Higher Gaussian Moments — Completing Wick's Theorem

There is a subtle point left before Wick's theorem is proved—what if more than two of the phis have the same momentum? If it's an odd number, the integral is zero; negative values cancel with the positive values. But if the number is even, the integral is positive. The previous demonstration assumed that the phis would only match up in pairs.

But the theorem is correct even when arbitrarily many of the φ are equal, and this is a notable property of Gaussian integration:

$$I = \int e^{-ax^2/2}dx = \sqrt{\frac{2\pi}{a}}$$

$$\frac{\partial^n}{\partial a^n}I = \int \frac{x^{2n}}{2^n}e^{-ax^2/2}dx = \frac{1\cdot 3\cdot 5\ldots\cdot(2n-1)}{2\cdot 2\cdot 2\ldots\quad\cdot 2}\sqrt{2\pi}\,a^{-\frac{2n+1}{2}}$$

Dividing by I,

$$\left\langle x^{2n}\right\rangle = \frac{\int x^{2n} e^{-ax^2/2}}{\int e^{-ax^2/2}} = 1\cdot 3\cdot 5\ldots\cdot(2n-1)\frac{1}{a^n}$$

$$\left\langle x^{2}\right\rangle = \frac{1}{a}$$

If Wick's theorem were correct, the higher moments would be given by all possible pairings of a list of $2n$ different x:

$$\left\langle x_1 x_2 x_3 \cdots x_{2n}\right\rangle$$

where the x are all the same variable, the index is just to keep track of the number of ways to pair them. The first x can be paired with $2n-1$ others, leaving $2n-2$ The next unpaired x can be paired with $2n-3$ different x leaving $2n-4$, and so on. This means that Wick's theorem, uncorrected, says that the expectation value of x^{2n} should be:

$$\left\langle x^{2n}\right\rangle = (2n-1)\cdot(2n-3)\ldots\cdot 5\cdot 3\cdot 1\left\langle x^2\right\rangle^n$$

and this is in fact the correct answer. So Wick's theorem holds no matter how many of the momenta of the internal variables coincide.

Interaction

Interactions are represented by higher order contributions, since quadratic contributions are always Gaussian. The simplest interaction is the quartic self-interaction, with an action:

$$S = \int \partial^\mu \phi \partial_\mu \phi + \frac{\lambda}{4!}\phi^4.$$

The reason for the combinatorial factor 4! will be clear soon. Writing the action in terms of the lattice (or continuum) Fourier modes:

$$S = \int_k k^2 |\phi(k)|^2 + \frac{\lambda}{4!}\int_{k_1k_2k_3k_4} \phi(k_1)\phi(k_2)\phi(k_3)\phi(k_4)\delta(k_1+k_2+k_3+k_4) = S_F + X.$$

Where S_F is the free action, whose correlation functions are given by Wick's theorem. The exponential of S in the path integral can be expanded in powers of λ, giving a series of corrections to the free action.

$$e^{-S} = e^{-S_F}\left(1 + X + \frac{1}{2!}XX + \frac{1}{3!}XXX + \cdots\right)$$

The path integral for the interacting action is then a power series of corrections to the free action. The term represented by X should be thought of as four half-lines, one for each factor of $\varphi(k)$. The half-lines meet at a vertex, which contributes a delta-function that ensures that the sum of the momenta are all equal.

To compute a correlation function in the interacting theory, there is a contribution from the X terms now. For example, the path-integral for the four-field correlator:

$$\langle \phi(k_1)\phi(k_2)\phi(k_3)\phi(k_4) \rangle = \frac{\int e^{-S}\phi(k_1)\phi(k_2)\phi(k_3)\phi(k_4)D\phi}{Z}$$

which in the free field was only nonzero when the momenta k were equal in pairs, is now nonzero for all values of k. The momenta of the insertions $\varphi(k_i)$ can now match up with the momenta of the Xs in the expansion. The insertions should also be thought of as half-lines, four in this case, which carry a momentum k, but one that is not integrated.

The lowest-order contribution comes from the first nontrivial term $e^{-S_F}X$ in the Taylor expansion of the action. Wick's theorem requires that the momenta in the X half-lines, the $\varphi(k)$ factors in X, should match up with the momenta of the external half-lines in pairs. The new contribution is equal to:

$$\lambda \frac{1}{k_1^2}\frac{1}{k_2^2}\frac{1}{k_3^2}\frac{1}{k_4^2}.$$

The 4! inside X is canceled because there are exactly 4! ways to match the half-lines in X to the external half-lines. Each of these different ways of matching the half-lines together in pairs contributes exactly once, regardless of the values of $k_{1,2,3,4}$, by Wick's theorem.

Feynman Diagrams

The expansion of the action in powers of X gives a series of terms with progressively higher number of Xs. The contribution from the term with exactly n Xs is called nth order.

The nth order terms has:

1. $4n$ internal half-lines, which are the factors of $\varphi(k)$ from the Xs. These all end on a vertex, and are integrated over all possible k.
2. external half-lines, which are the come from the $\varphi(k)$ insertions in the integral.

By Wick's theorem, each pair of half-lines must be paired together to make a *line*, and this line gives a factor of

$$\frac{\delta(k_1 + k_2)}{k_1^2}$$

which multiplies the contribution. This means that the two half-lines that make a line are forced to have equal and opposite momentum. The line itself should be labelled by an arrow, drawn parallel to the line, and labeled by the momentum in the line k. The half-line at the tail end of the arrow carries momentum k, while the half-line at the head-end carries momentum $-k$. If one of the two half-lines is external, this kills the integral over the internal k, since it forces the internal k to be equal to the external k. If both are internal, the integral over k remains.

The diagrams that are formed by linking the half-lines in the Xs with the external half-lines, representing insertions, are the Feynman diagrams of this theory. Each line carries a factor of $1/k^2$, the propagator, and either goes from vertex to vertex, or ends at an insertion. If it is internal, it is integrated over. At each vertex, the total incoming k is equal to the total outgoing k.

The number of ways of making a diagram by joining half-lines into lines almost completely cancels the factorial factors coming from the Taylor series of the exponential and the 4! at each vertex.

Loop Order

A forest diagram is one where all the internal lines have momentum that is completely determined by the external lines and the condition that the incoming and outgoing momentum are equal at each vertex. The contribution of these diagrams is a product of propagators, without any integration. A tree diagram is a connected forest diagram.

An example of a tree diagram is the one where each of four external lines end on an X. Another is when three external lines end on an X, and the remaining half-line joins up with another X, and the remaining half-lines of this X run off to external lines. These are all also forest diagrams (as every tree is a forest); an example of a forest that is not a tree is when eight external lines end on two Xs.

It is easy to verify that in all these cases, the momenta on all the internal lines is determined by the external momenta and the condition of momentum conservation in each vertex.

A diagram that is not a forest diagram is called a *loop* diagram, and an example is one where two lines of an X are joined to external lines, while the remaining two lines are joined to each other. The two lines joined to each other can have any momentum at all, since they both enter and leave the same vertex. A more complicated example is one where two Xs are joined to each other by matching the legs one to the other. This diagram has no external lines at all.

The reason loop diagrams are called loop diagrams is because the number of k-integrals that are left undetermined by momentum conservation is equal to the number of independent closed loops in the diagram, where independent loops are counted as in homology theory. The homology is real-valued (actually R^d valued), the value associated with each line is the momentum. The boundary operator takes each line to the sum of the end-vertices with a positive sign at the head and a negative sign at the tail. The condition that the momentum is conserved is exactly the condition that the boundary of the k-valued weighted graph is zero.

A set of valid k-values can be arbitrarily redefined whenever there is a closed loop. A closed loop is a cyclical path of adjacent vertices that never revisits the same vertex. Such a cycle can be thought of as the boundary of a hypothetical 2-cell. The k-labellings of a graph that conserve momentum (i.e. which has zero boundary) up to redefinitions of k (i.e. up to boundaries of 2-cells) define the first homology of a graph. The number of independent momenta that are not determined is then equal to the number of independent homology loops. For many graphs, this is equal to the number of loops as counted in the most intuitive way.

Symmetry Factors

The number of ways to form a given Feynman diagram by joining together half-lines is large, and by Wick's theorem, each way of pairing up the half-lines contributes equally. Often, this completely cancels the factorials in the denominator of each term, but the cancellation is sometimes incomplete.

The uncancelled denominator is called the *symmetry factor* of the diagram. The contribution of each diagram to the correlation function must be divided by its symmetry factor.

For example, consider the Feynman diagram formed from two external lines joined to one X, and the remaining two half-lines in the X joined to each other. There are 4×3 ways to join the external half-lines to the X, and then there is only one way to join the two remaining lines to each other. The X comes divided by $4! = 4 \times 3 \times 2$, but the number of ways to link up the X half lines to make the diagram is only 4×3, so the contribution of this diagram is divided by two.

For another example, consider the diagram formed by joining all the half-lines of one X to all the half-lines of another X. This diagram is called a *vacuum bubble*, because it does not link up to any external lines. There are 4! ways to form this diagram, but the denominator includes a 2! (from the expansion of the exponential, there are two Xs) and two factors of 4! The contribution is multiplied by,

$$\frac{4!}{2 \times 4! \times 4!} = \frac{1}{48}$$

Another example is the Feynman diagram formed from two Xs where each X links up to two external lines, and the remaining two half-lines of each X are joined to each other. The number of ways to link an X to two external lines is 4×3, and either X could link up to either pair, giving an additional factor of 2. The remaining two half-lines in the two Xs can be linked to each other in two ways, so that the total number of ways to form the diagram is $4 \times 3 \times 4 \times 3 \times 2 \times 2$, while the denominator is $4! \times 4! \times 2!$. The total symmetry factor is 2, and the contribution of this diagram is divided by 2.

The symmetry factor theorem gives the symmetry factor for a general diagram: the contribution of each Feynman diagram must be divided by the order of its group of automorphisms, the number of symmetries that it has.

An automorphism of a Feynman graph is a permutation M of the lines and a permutation N of the vertices with the following properties:

1. If a line l goes from vertex v to vertex v', then $M(l)$ goes from $N(v)$ to $N(v)$ If the line is undirected, as it is for a real scalar field, then $M(l)$ can go from $N(v)$ to $N(v)$ too.
2. If a line l ends on an external line, $M(l)$ ends on the same external line.
3. If there are different types of lines, $M(l)$ should preserve the type.

This theorem has an interpretation in terms of particle-paths: when identical particles are present, the integral over all intermediate particles must not double-count states that differ only by interchanging identical particles.

Proof: To prove this theorem, label all the internal and external lines of a diagram with a unique name. Then form the diagram by linking a half-line to a name and then to the other half line.

Now count the number of ways to form the named diagram. Each permutation of the *X*s gives a different pattern of linking names to half-lines, and this is a factor of $n!$. Each permutation of the half-lines in a single X gives a factor of $4!$. So a named diagram can be formed in exactly as many ways as the denominator of the Feynman expansion.

But the number of unnamed diagrams is smaller than the number of named diagram by the order of the automorphism group of the graph.

Connected Diagrams: Linked-cluster Theorem

Roughly speaking, a Feynman diagram is called *connected* if all vertices and propagator lines are linked by a sequence of vertices and propagators of the diagram itself. If one views it as an undirected graph it is connected. The remarkable relevance of such diagrams in QFTs is due to the fact that they are sufficient to determine the quantum partition function $Z[J]$. More precisely, connected Feynman diagrams determine,

$$iW[J] \equiv \ln Z[J].$$

one should recall that:

$$Z[J] \propto \sum_k D_k$$

with D_k constructed from some (arbitrary) Feynman diagram that can be thought to consist of several connected components C_i. If one encounters n_i (identical) copies of a component C_i within the Feynman diagram D_k one has to include a *symmetry factor* $n_i!$. However, in the end each contribution of a Feynman diagram D_k to the partition function has the generic form,

$$\prod_i \frac{C_i^{n_i}}{n_i!}$$

where i labels the (infinitely) many connected Feynman diagrams possible.

A scheme to successively create such contributions from the D_k to $Z[J]$ is obtained by,

$$\left(\frac{1}{0!}+\frac{C_1}{1!}+\frac{C_1^2}{2!}+\cdots\right)\left(1+C_2+\frac{1}{2}C_2^2+\cdots\right)$$

and therefore yields,

$$Z[J] \propto \prod_i \sum_{n_i=0}^{\infty} \frac{C_i^{n_i}}{n_i!} = \exp \sum_i C_i \propto \exp W[J].$$

To establish the *normalization* $Z_0 = \exp W[0] = 1$ one simply calculates all connected *vacuum diagrams*, i.e., the diagrams without any *sources J*.

Vacuum Bubbles

An immediate consequence of the linked-cluster theorem is that all vacuum bubbles, diagrams without external lines, cancel when calculating correlation functions. A correlation function is given by a ratio of path-integrals:

$$\langle \phi_1(x_1)\cdots\phi_n(x_n)\rangle = \frac{\int e^{-S}\phi_1(x_1)\cdots\phi_n(x_n)D\phi}{\int e^{-S}D\phi}.$$

The top is the sum over all Feynman diagrams, including disconnected diagrams that do not link up to external lines at all. In terms of the connected diagrams, the numerator includes the same contributions of vacuum bubbles as the denominator:

$$\int e^{-S}\phi_1(x_1)\cdots\phi_n(x_n)D\phi = \left(\sum E_i\right)\left(\exp\left(\sum_i C_i\right)\right).$$

Where the sum over E diagrams includes only those diagrams each of whose connected components end on at least one external line. The vacuum bubbles are the same whatever the external lines, and give an overall multiplicative factor. The denominator is the sum over all vacuum bubbles, and dividing gets rid of the second factor.

The vacuum bubbles then are only useful for determining Z itself, which from the definition of the path integral is equal to:

$$Z = \int e^{-S}D\phi = e^{-HT} = e^{-\rho V}$$

where ρ is the energy density in the vacuum. Each vacuum bubble contains a factor of $\delta(k)$ zeroing the total k at each vertex, and when there are no external lines, this contains a factor of $\delta(0)$, because the momentum conservation is over-enforced. In finite volume, this factor can be identified as the total volume of space time. Dividing by the volume, the remaining integral for the vacuum bubble has an interpretation: it is a contribution to the energy density of the vacuum.

Correlation functions are the sum of the connected Feynman diagrams, but the formalism treats the connected and disconnected diagrams differently. Internal lines end on vertices, while external lines go off to insertions. Introducing *sources* unifies the formalism, by making new vertices where one line can end.

Sources are external fields, fields that contribute to the action, but are not dynamical variables. A scalar field source is another scalar field h that contributes a term to the (Lorentz) Lagrangian:

$$\int h(x)\phi(x)d^dx = \int h(k)\phi(k)d^dk$$

In the Feynman expansion, this contributes H terms with one half-line ending on a vertex. Lines in a Feynman diagram can now end either on an X vertex, or on an H vertex, and only one line enters an H vertex. The Feynman rule for an H vertex is that a line from an H with momentum k gets a factor of $h(k)$.

The sum of the connected diagrams in the presence of sources includes a term for each connected diagram in the absence of sources, except now the diagrams can end on the source. Traditionally, a source is represented by a little "×" with one line extending out, exactly as an insertion.

$$\log\left(Z[h]\right)=\sum_{n,C} h(k_1)h(k_2)\cdots h(k_n)C(k_1,\cdots,k_n)$$

where $C(k_1,\ldots,k_n)$ is the connected diagram with n external lines carrying momentum as indicated. The sum is over all connected diagrams, as before.

The field h is not dynamical, which means that there is no path integral over h: h is just a parameter in the Lagrangian, which varies from point to point. The path integral for the field is:

$$Z[h]=\int e^{iS+i\int h\phi}\,D\phi$$

and it is a function of the values of h at every point. One way to interpret this expression is that it is taking the Fourier transform in field space. If there is a probability density on R^n, the Fourier transform of the probability density is:

$$\int \rho(y)e^{iky}\,d^n y=\left\langle e^{iky}\right\rangle=\left\langle \prod_{i=1}^{n} e^{ih_i y_i}\right\rangle$$

The Fourier transform is the expectation of an oscillatory exponential. The path integral in the presence of a source $h(x)$ is:

$$Z[h]=\int e^{iS}e^{i\int_x h(x)\phi(x)}\,D\phi=\left\langle e^{ih\phi}\right\rangle$$

which, on a lattice, is the product of an oscillatory exponential for each field value:

$$\left\langle \prod_x e^{ih_x\phi_x}\right\rangle$$

The Fourier transform of a delta-function is a constant, which gives a formal expression for a delta function:

$$\delta(x-y)=\int e^{ik(x-y)}\,dk$$

This tells you what a field delta function looks like in a path-integral. For two scalar fields φ and η,

$$\delta(\phi-\eta)=\int e^{ih(x)\left(\phi(x)-\eta(x)\right)d^dx}\,Dh,$$

which integrates over the Fourier transform coordinate, over h. This expression is useful for formally changing field coordinates in the path integral, much as a delta function is used to change coordinates in an ordinary multi-dimensional integral.

The partition function is now a function of the field h, and the physical partition function is the value when h is the zero function:

The correlation functions are derivatives of the path integral with respect to the source:

$$\langle\phi(x)\rangle = \frac{1}{Z}\frac{\partial}{\partial h(x)}Z[h] = \frac{\partial}{\partial h(x)}\log(Z[h]).$$

In Euclidean space, source contributions to the action can still appear with a factor of i, so that they still do a Fourier transform.

Spin $\frac{1}{2}$: "photons" and "ghosts".

Spin $\frac{1}{2}$: Grassmann integrals.

The field path integral can be extended to the Fermi case, but only if the notion of integration is expanded. A Grassmann integral of a free Fermi field is a high-dimensional determinant or Pfaffian, which defines the new type of Gaussian integration appropriate for Fermi fields.

The two fundamental formulas of Grassmann integration are:

$$\int e^{M_{ij}\bar{\psi}^i\psi^j}\, D\bar{\psi}\, D\psi = \mathrm{Det}(M),$$

where M is an arbitrary matrix and ψ, $\bar{\psi}$ are independent Grassmann variables for each index i, and

$$\int e^{\frac{1}{2}A_{ij}\psi^i\psi^j}\, D\psi = \mathrm{Pfaff}(A),$$

where A is an antisymmetric matrix, ψ is a collection of Grassmann variables, and the $\frac{1}{2}$ is to prevent double-counting (since $\psi^i\psi^j = -\psi^j\psi^i$).

In matrix notation, where $\bar{\psi}$ and $\bar{\eta}$ are Grassmann-valued row vectors, η and ψ are Grassmann-valued column vectors, and M is a real-valued matrix:

$$Z = \int e^{\bar{\psi}M\psi + \bar{\eta}\psi + \bar{\psi}\eta}\, D\bar{\psi}\, D\psi = \int e^{(\bar{\psi}+\bar{\eta}M^{-1})M(\psi+M^{-1}\eta) - \bar{\eta}M^{-1}\eta}\, D\bar{\psi}\, D\psi = \mathrm{Det}(M)e^{-\bar{\eta}M^{-1}\eta},$$

where the last equality is a consequence of the translation invariance of the Grassmann integral. The Grassmann variables η are external sources for ψ, and differentiating with respect to $\bar{\eta}$ pulls down factors of $\bar{\psi}$.

$$\langle\bar{\psi}\psi\rangle = \frac{1}{Z}\frac{\partial}{\partial\eta}\frac{\partial}{\partial\bar{\eta}}Z\,|_{\eta=\bar{\eta}=0} = M^{-1}$$

again, in a schematic matrix notation. The meaning of the formula above is that the derivative with respect to the appropriate component of η and $\bar{\eta}$ gives the matrix element of M^{-1}. This is exactly analogous to the bosonic path integration formula for a Gaussian integral of a complex bosonic field:

$$\int e^{\phi^* M\phi + h^*\phi + \phi^* h} D\phi^* D\phi = \frac{e^{h^* M^{-1} h}}{\mathrm{Det}(M)}$$

$$\langle \phi^* \phi \rangle = \frac{1}{Z} \frac{\partial}{\partial h} \frac{\partial}{\partial h^*} Z \,|_{h=h^*=0} = M^{-1}.$$

So that the propagator is the inverse of the matrix in the quadratic part of the action in both the Bose and Fermi case.

For real Grassmann fields, for Majorana fermions, the path integral a Pfaffian times a source quadratic form, and the formulas give the square root of the determinant, just as they do for real Bosonic fields. The propagator is still the inverse of the quadratic part.

The free Dirac Lagrangian:

$$\int \bar{\psi}\left(\gamma^\mu \partial_\mu - m\right)\psi$$

formally gives the equations of motion and the anticommutation relations of the Dirac field, just as the Klein Gordon Lagrangian in an ordinary path integral gives the equations of motion and commutation relations of the scalar field. By using the spatial Fourier transform of the Dirac field as a new basis for the Grassmann algebra, the quadratic part of the Dirac action becomes simple to invert:

$$S = \int_k \bar{\psi}\left(i\gamma^\mu k_\mu - m\right)\psi.$$

The propagator is the inverse of the matrix M linking $\psi(k)$ and $\bar{\psi}(k)$, since different values of k do not mix together.

$$\langle \bar{\psi}(k')\psi(k) \rangle = \delta(k+k')\frac{1}{\gamma\cdot k - m} = \delta(k+k')\frac{\gamma\cdot k + m}{k^2 - m^2}$$

The analog of Wick's theorem matches ψ and $\bar{\psi}$ in pairs:

$$\langle \bar{\psi}(k_1)\bar{\psi}(k_2)\cdots\bar{\psi}(k_n)\psi(k_1')\cdots\psi(k_n) \rangle = \sum_{\text{pairings}} (-1)^S \prod_{\text{pairs } i,j} \delta\left(k_i - k_j\right)\frac{1}{\gamma\cdot k_i - m}$$

where S is the sign of the permutation that reorders the sequence of ψ and ψ to put the ones that are paired up to make the delta-functions next to each other, with the ψ coming right before the ψ. Since a ψ, ψ pair is a commuting element of the Grassmann algebra, it doesn't matter what order the pairs are in. If more than one ψ, ψ pair have the same k, the integral is zero, and it is easy to check that the sum over pairings gives zero in this case (there are always an even number of them). This is the Grassmann analog of the higher Gaussian moments that completed the Bosonic Wick's theorem earlier.

The rules for spin-1/2 Dirac particles are as follows: The propagator is the inverse of the Dirac operator, the lines have arrows just as for a complex scalar field, and the diagram acquires an overall factor of −1 for each closed Fermi loop. If there are an odd number of Fermi loops, the diagram changes sign. Historically, the −1 rule was very difficult for Feynman to discover. He discovered it after a long process of trial and error, since he lacked a proper theory of Grassmann integration.

The rule follows from the observation that the number of Fermi lines at a vertex is always even. Each term in the Lagrangian must always be Bosonic. A Fermi loop is counted by following Fermionic lines until one comes back to the starting point, then removing those lines from the diagram. Repeating this process eventually erases all the Fermionic lines: this is the Euler algorithm to 2-color a graph, which works whenever each vertex has even degree. Note that the number of steps in the Euler algorithm is only equal to the number of independent Fermionic homology cycles in the common special case that all terms in the Lagrangian are exactly quadratic in the Fermi fields, so that each vertex has exactly two Fermionic lines. When there are four-Fermi interactions (like in the Fermi effective theory of the weak nuclear interactions) there are more k-integrals than Fermi loops. In this case, the counting rule should apply the Euler algorithm by pairing up the Fermi lines at each vertex into pairs that together form a bosonic factor of the term in the Lagrangian, and when entering a vertex by one line, the algorithm should always leave with the partner line.

To clarify and prove the rule, consider a Feynman diagram formed from vertices, terms in the Lagrangian, with Fermion fields. The full term is Bosonic, it is a commuting element of the Grassmann algebra, so the order in which the vertices appear is not important. The Fermi lines are linked into loops, and when traversing the loop, one can reorder the vertex terms one after the other as one goes around without any sign cost. The exception is when you return to the starting point, and the final half-line must be joined with the unlinked first half-line. This requires one permutation to move the last ψ to go in front of the first ψ, and this gives the sign.

This rule is the only visible effect of the exclusion principle in internal lines. When there are external lines, the amplitudes are antisymmetric when two Fermi insertions for identical particles are interchanged. This is automatic in the source formalism, because the sources for Fermi fields are themselves Grassmann valued.

Spin 1: Photons

The naive propagator for photons is infinite, since the Lagrangian for the A-field is:

$$S = \int \tfrac{1}{4} F^{\mu\nu} F_{\mu\nu} = \int -\tfrac{1}{2}\left(\partial^{\mu} A_{\nu} \partial_{\mu} A^{\nu} - \partial^{\mu} A_{\mu} \partial_{\nu} A^{\nu}\right).$$

The quadratic form defining the propagator is non-invertible. The reason is the gauge invariance of the field; adding a gradient to A does not change the physics.

To fix this problem, one needs to fix a gauge. The most convenient way is to demand that the divergence of A is some function f, whose value is random from point to point. It does no harm to integrate over the values of f, since it only determines the choice of gauge. This procedure inserts the following factor into the path integral for A:

$$\int \delta\left(\partial_{\mu} A^{\mu} - f\right) e^{-\frac{f^2}{2}} Df.$$

The first factor, the delta function, fixes the gauge. The second factor sums over different values of f that are inequivalent gauge fixings. This is simply,

$$e^{-\frac{(\partial_\mu A_\mu)^2}{2}}.$$

The additional contribution from gauge-fixing cancels the second half of the free Lagrangian, giving the Feynman Lagrangian:

$$S = \int \partial^\mu A^\nu \partial_\mu A_\nu$$

which is just like four independent free scalar fields, one for each component of A. The Feynman propagator is:

$$\langle A_\mu(k) A_\nu(k') \rangle = \delta(k+k') \frac{g_{\mu\nu}}{k^2}.$$

The one difference is that the sign of one propagator is wrong in the Lorentz case: the timelike component has an opposite sign propagator. This means that these particle states have negative norm—they are not physical states. In the case of photons, it is easy to show by diagram methods that these states are not physical—their contribution cancels with longitudinal photons to only leave two physical photon polarization contributions for any value of k.

If the averaging over f is done with a coefficient different from $\frac{1}{2}$, the two terms don't cancel completely. This gives a covariant Lagrangian with a coefficient λ, which does not affect anything:

$$S = \int \tfrac{1}{2}\left(\partial^\mu A^\nu \partial_\mu A_\nu - \lambda \left(\partial_\mu A^\mu\right)^2\right)$$

and the covariant propagator for QED is:

$$\langle A_\mu(k) A_\nu(k') \rangle = \delta(k+k') \frac{g_{\mu\nu} - \lambda \frac{k_\mu k_\nu}{k^2}}{k^2}.$$

Spin 2: Non-Abelian Ghosts

To find the Feynman rules for non-Abelian gauge fields, the procedure that performs the gauge fixing must be carefully corrected to account for a change of variables in the path-integral.

The gauge fixing factor has an extra determinant from popping the delta function:

$$\delta\left(\partial_\mu A_\mu - f\right) e^{-\frac{f^2}{2}} \det M$$

To find the form of the determinant, consider first a simple two-dimensional integral of a function f that depends only on r, not on the angle θ. Inserting an integral over θ:

$$\int f(r) dx dy = \int f(r) \int d\theta \delta(y) \left| \frac{dy}{d\theta} \right| dx dy$$

The derivative-factor ensures that popping the delta function in θ removes the integral. Exchanging the order of integration,

$$\int f(r)dxdy = \int d\theta \int f(r)\delta(y)\left|\frac{dy}{d\theta}\right| dxdy$$

but now the delta-function can be popped in y,

$$\int f(r)dxdy = \int d\theta_0 \int f(x)\left|\frac{dy}{d\theta}\right| dx.$$

The integral over θ just gives an overall factor of 2π, while the rate of change of y with a change in θ is just x, so this exercise reproduces the standard formula for polar integration of a radial function:

$$\int f(r)dxdy = 2\pi \int f(x)xdx$$

In the path-integral for a nonabelian gauge field, the analogous manipulation is:

$$\int DA \int \delta\big(F(A)\big)\det\left(\frac{\partial F}{\partial G}\right) DGe^{iS} = \int DG \int \delta\big(F(A)\big)\det\left(\frac{\partial F}{\partial G}\right) e^{iS}$$

The factor in front is the volume of the gauge group, and it contributes a constant, which can be discarded. The remaining integral is over the gauge fixed action.

$$\int \det\left(\frac{\partial F}{\partial G}\right) e^{iS_{GF}}\, DA$$

To get a covariant gauge, the gauge fixing condition is the same as in the Abelian case:

$$\partial_\mu A^\mu = f,$$

Whose variation under an infinitesimal gauge transformation is given by:

$$\partial_\mu D_\mu \alpha,$$

where α is the adjoint valued element of the Lie algebra at every point that performs the infinitesimal gauge transformation. This adds the Faddeev Popov determinant to the action:

$$\det\left(\partial_\mu D_\mu\right)$$

which can be rewritten as a Grassmann integral by introducing ghost fields:

$$\int e^{\bar{\eta}\partial_\mu D^\mu \eta}\, D\bar{\eta} D\eta$$

The determinant is independent of f, so the path-integral over f can give the Feynman propagator (or a covariant propagator) by choosing the measure for f as in the abelian case. The full gauge fixed action is then the Yang Mills action in Feynman gauge with an additional ghost action:

$$S=\int \mathrm{Tr}\partial_{\mu}A_{\nu}\partial^{\mu}A^{\nu}+f^{i}_{jk}\partial^{\nu}A^{\mu}_{i}A^{j}_{\mu}A^{k}_{\nu}+f^{i}_{jr}f^{r}_{kl}A_{i}A_{j}A^{k}A^{l}+\mathrm{Tr}\partial_{\mu}\bar{\eta}\partial^{\mu}\eta+\bar{\eta}A_{j}\eta S$$

The diagrams are derived from this action. The propagator for the spin-1 fields has the usual Feynman form. There are vertices of degree 3 with momentum factors whose couplings are the structure constants, and vertices of degree 4 whose couplings are products of structure constants. There are additional ghost loops, which cancel out timelike and longitudinal states in A loops.

In the Abelian case, the determinant for covariant gauges does not depend on A, so the ghosts do not contribute to the connected diagrams.

Particle-path Representation

Feynman diagrams were originally discovered by Feynman, by trial and error, as a way to represent the contribution to the S-matrix from different classes of particle trajectories.

Schwinger Representation

The Euclidean scalar propagator has a suggestive representation:

$$\frac{1}{p^2+m^2}=\int_0^\infty e^{-\tau\left(p^2+m^2\right)}d\tau$$

The meaning of this identity (which is an elementary integration) is made clearer by Fourier transforming to real space.

$$\Delta(x)=\int_0^\infty d\tau e^{-m^2\tau}\frac{1}{(4\pi\tau)^{d/2}}e^{\frac{-x^2}{4\tau}}$$

The contribution at any one value of τ to the propagator is a Gaussian of width $\sqrt{\tau}$ The total propagation function from 0 to x is a weighted sum over all proper times τ of a normalized Gaussian, the probability of ending up at x after a random walk of time $\sqrt{\tau}$.

The path-integral representation for the propagator is then:

$$\Delta(x)=\int_0^\infty d\tau\int DXe^{-\int_0^\tau\left(\frac{\dot{x}^2}{2}+m^2\right)d\tau'}$$

which is a path-integral rewrite of the Schwinger representation.

The Schwinger representation is both useful for making manifest the particle aspect of the propagator, and for symmetrizing denominators of loop diagrams.

Combining Denominators

The Schwinger representation has an immediate practical application to loop diagrams. For example, for the diagram in the φ^4 theory formed by joining two *x*s together in two half-lines, and making the remaining lines external, the integral over the internal propagators in the loop is:

$$\int_k \frac{1}{k^2+m^2}\frac{1}{(k+p)^2+m^2}.$$

Here one line carries momentum k and the other $k + p$. The asymmetry can be fixed by putting everything in the Schwinger representation.

$$\int_{t,t'} e^{-t(k^2+m^2)-t'\left((k+p)^2+m^2\right)}dt\,dt'.$$

Now the exponent mostly depends on $t+t'$,

$$\int_{t,t'} e^{-(t+t')(k^2+m^2)-t'2p\cdot k-t'p^2},$$

except for the asymmetrical little bit. Defining the variable $u = t + t'$ and $v=\frac{t'}{u'}$ variable u goes from 0 to ∞, while v goes from 0 to 1. The variable u is the total proper time for the loop, while v parametrizes the fraction of the proper time on the top of the loop versus the bottom.

The Jacobian for this transformation of variables is easy to work out from the identities:

$$d(uv)=dt' \quad du=dt+dt',$$

and "wedging" gives,

$$u\,du\wedge dv=dt\wedge dt'.$$

This allows the u integral to be evaluated explicitly:

$$\int_{u,v} ue^{-u\left(k^2+m^2+v2p\cdot k+vp^2\right)}=\int\frac{1}{\left(k^2+m^2+v2p\cdot k-vp^2\right)^2}$$

leaving only the v-integral. This method, invented by Schwinger but usually attributed to Feynman, is called *combining denominator*. Abstractly, it is the elementary identity:

$$\frac{1}{AB}=\int_0^1\frac{1}{\left(vA+(1-v)B\right)^2}dv$$

But this form does not provide the physical motivation for introducing $v; v$ is the proportion of proper time on one of the legs of the loop.

Once the denominators are combined, a shift in $k' = k + vp$ symmetrizes everything:

$$\int_0^1\int \frac{1}{\left(k^2+m^2+2vp\cdot k+vp^2\right)^2}dk\,dv=\int_0^1\int \frac{1}{\left(k'^2+m^2+v(1-v)p^2\right)^2}dk'\,dv$$

This form shows that the moment that p^2 is more negative than four times the mass of the particle in the loop, which happens in a physical region of Lorentz space, the integral has a cut. This is exactly when the external momentum can create physical particles.

When the loop has more vertices, there are more denominators to combine:

$$\int dk \frac{1}{k^2+m^2}\frac{1}{(k+p_1)^2+m^2}\cdots\frac{1}{(k+p_n)^2+m^2}$$

The general rule follows from the Schwinger prescription for $n + 1$ denominators:

$$\frac{1}{D_0 D_1\cdots D_n}=\int_0^\infty\cdots\int_0^\infty e^{-u_0D_0\cdots-u_nD_n}\,du_0\cdots du_n.$$

The integral over the Schwinger parameters u_i can be split up as before into an integral over the total proper time $u=u_0+u_1\ldots+u_n$ and an integral over the fraction of the proper time in all but the first segment of the loop $v_i=\frac{u_i}{u}$ for $i\in\ 1,2,\ldots,n$ The v are positive and add up to less than 1, so that the v integral is over an n-dimensional simplex.

The Jacobian for the coordinate transformation can be worked out as before:

$$du = du_0 + du_1 \cdots + du_n$$
$$d(uv_i) = du_i.$$

Wedging all these equations together, one obtains,

$$u^n\,du\wedge dv_1\wedge dv_2\cdots\wedge dv_n = du_0\wedge du_1\cdots\wedge du_n.$$

This gives the integral:

$$\int_0^\infty\int_{\text{simplex}} u^n e^{-u(v_0D_0+v_1D_1+v_2D_2\cdots+v_nD_n)}\,dv_1\cdots dv_n\,du,$$

where the simplex is the region defined by the conditions,

$$v_i > 0 \quad\text{and}\quad \sum_{i=1}^n v_i < 1$$

as well as,

$$v_0 = 1-\sum_{i=1}^n v_i.$$

Performing the u integral gives the general prescription for combining denominators:

$$\frac{1}{D_0 \cdots D_n} = n! \int_{\text{simplex}} \frac{1}{(v_0 D_0 + v_1 D_1 \cdots + v_n D_n)^{n+1}} dv_1\, dv_2 \cdots dv_n$$

Since the numerator of the integrand is not involved, the same prescription works for any loop, no matter what the spins are carried by the legs. The interpretation of the parameters v_i is that they are the fraction of the total proper time spent on each leg.

Scattering

The correlation functions of a quantum field theory describe the scattering of particles. The definition of "particle" in relativistic field theory is not self-evident, because if you try to determine the position so that the uncertainty is less than the compton wavelength, the uncertainty in energy is large enough to produce more particles and antiparticles of the same type from the vacuum. This means that the notion of a single-particle state is to some extent incompatible with the notion of an object localized in space.

In the 1930s, Wigner gave a mathematical definition for single-particle states: they are a collection of states that form an irreducible representation of the Poincaré group. Single particle states describe an object with a finite mass, a well defined momentum, and a spin. This definition is fine for protons and neutrons, electrons and photons, but it excludes quarks, which are permanently confined, so the modern point of view is more accommodating: a particle is anything whose interaction can be described in terms of Feynman diagrams, which have an interpretation as a sum over particle trajectories.

A field operator can act to produce a one-particle state from the vacuum, which means that the field operator $\varphi(x)$ produces a superposition of Wigner particle states. In the free field theory, the field produces one particle states only. But when there are interactions, the field operator can also produce 3-particle, 5-particle (if there is no +/− symmetry also 2, 4, 6 particle) states too. To compute the scattering amplitude for single particle states only requires a careful limit, sending the fields to infinity and integrating over space to get rid of the higher-order corrections.

The relation between scattering and correlation functions is the LSZ-theorem: The scattering amplitude for n particles to go to m particles in a scattering event is the given by the sum of the Feynman diagrams that go into the correlation function for $n + m$ field insertions, leaving out the propagators for the external legs.

For example, for the $\lambda\varphi^4$ interaction of the previous section, the order λ contribution to the (Lorentz) correlation function is:

$$\langle \phi(k_1)\phi(k_2)\phi(k_3)\phi(k_4) \rangle = \frac{i}{k_1^2}\frac{i}{k_2^2}\frac{i}{k_3^2}\frac{i}{k_4^2} i\lambda$$

Stripping off the external propagators, that is, removing the factors of $\frac{i}{k^2}$, gives the invariant scattering amplitude M:

$$M = i\lambda$$

which is a constant, independent of the incoming and outgoing momentum. The interpretation of the scattering amplitude is that the sum of $|M|^2$ over all possible final states is the probability for the scattering event. The normalization of the single-particle states must be chosen carefully, however, to ensure that M is a relativistic invariant.

Non-relativistic single particle states are labeled by the momentum k, and they are chosen to have the same norm at every value of k. This is because the nonrelativistic unit operator on single particle states is:

$$\int dk|\, k\rangle\langle k\,|.$$

In relativity, the integral over the k-states for a particle of mass m integrates over a hyperbola in E,k space defined by the energy–momentum relation:

$$E^2 - k^2 = m^2.$$

If the integral weighs each k point equally, the measure is not Lorentz-invariant. The invariant measure integrates over all values of k and E, restricting to the hyperbola with a Lorentz-invariant delta function:

$$\int \delta(E^2 - k^2 - m^2)\,|\, E,k\rangle\langle E,k\,| dE\, dk = \int \frac{dk}{2E}\,|\, k\rangle\langle k\,|.$$

So the normalized k-states are different from the relativistically normalized k-states by a factor of

$$\sqrt{E} = \left(k^2 - m^2\right)^{\frac{1}{4}}.$$

The invariant amplitude M is then the probability amplitude for relativistically normalized incoming states to become relativistically normalized outgoing states.

For nonrelativistic values of k, the relativistic normalization is the same as the nonrelativistic normalization (up to a constant factor $\sqrt{m}$). In this limit, the φ^4 invariant scattering amplitude is still constant. The particles created by the field φ scatter in all directions with equal amplitude.

The nonrelativistic potential, which scatters in all directions with an equal amplitude (in the Born approximation), is one whose Fourier transform is constant—a delta-function potential. The lowest order scattering of the theory reveals the non-relativistic interpretation of this theory—it describes a collection of particles with a delta-function repulsion. Two such particles have an aversion to occupying the same point at the same time.

Nonperturbative Effects

Thinking of Feynman diagrams as a perturbation series, nonperturbative effects like tunneling do not show up, because any effect that goes to zero faster than any polynomial does not affect the Taylor series. Even bound states are absent, since at any finite order particles are only exchanged a finite number of times, and to make a bound state, the binding force must last forever.

But this point of view is misleading, because the diagrams not only describe scattering, but they also are a representation of the short-distance field theory correlations. They encode not only asymptotic processes like particle scattering, they also describe the multiplication rules for fields, the operator product expansion. Nonperturbative tunneling processes involve field configurations that on average get big when the coupling constant gets small, but each configuration is a coherent superposition of particles whose local interactions are described by Feynman diagrams. When the coupling is small, these become collective processes that involve large numbers of particles, but where the interactions between each of the particles is simple.

This means that nonperturbative effects show up asymptotically in resummations of infinite classes of diagrams, and these diagrams can be locally simple. The graphs determine the local equations of motion, while the allowed large-scale configurations describe non-perturbative physics. But because Feynman propagators are nonlocal in time, translating a field process to a coherent particle language is not completely intuitive, and has only been explicitly worked out in certain special cases. In the case of nonrelativistic bound states, the Bethe–Salpeter equation describes the class of diagrams to include to describe a relativistic atom. For quantum chromodynamics, the Shifman Vainshtein Zakharov sum rules describe non-perturbatively excited long-wavelength field modes in particle language, but only in a phenomenological way.

The number of Feynman diagrams at high orders of perturbation theory is very large, because there are as many diagrams as there are graphs with a given number of nodes. Nonperturbative effects leave a signature on the way in which the number of diagrams and resummations diverge at high order. It is only because non-perturbative effects appear in hidden form in diagrams that it was possible to analyze nonperturbative effects in string theory, where in many cases a Feynman description is the only one available.

Casimir's Trick

Casimir's trick, named after the Dutch physicist Hendrik Casimir, is a method for the simple calculation of spin-averaged squared matrix elements in quantum field theories.

General

A common expression in quantum field theories is the matrix element of the S matrix $\mathcal{M}$, which describes the transition from an initial state to a final state. This matrix element can be graphically represented using Feynman diagrams and translated into a rigorous mathematical expression. Are fermions, i.e. particles with a spin of $1/2$ involved, Dirac spinors, i.e. four-component vectors with additional spin indices, appear in the calculations of the matrix elements.

A lorentzin variant, scalar size, is the squared matrix element $|\mathcal{M}|^2 = \mathcal{M}^* \mathcal{M}$, which generally contains complex expressions from Dirac spinor products. However, if you are only interested in a matrix element averaged over all possible spin settings in the calculation, Casimir's trick can be used to convert the matrix element into a product of traces using Dirac matrices, which can be easily executed on the basis of Dirac algebra.

Details

Describe $u(k)$,$[\bar{v}(k)]$ the spinors for incoming [anti] particles in Feynman diagrams and $\bar{u}(k)[v(k)]$ Spinors for leaking [anti] particles, the following applies:

$$\sum_{s_1,s_2}\left[\bar{v}^{s_1}(k)\Gamma u^{s_2}(p)\right]\left[\bar{v}^{s_1}(k)\Gamma u^{s_2}(p)\right]^* = \mathrm{Tr}\left[\Gamma\left(\gamma^\alpha p_\alpha + m_1\right)\bar{\Gamma}\left(\gamma^\beta k_\beta - m_2\right)\right]$$

$$\sum_{s_1,s_2}\left[\bar{u}^{s_1}(k)\Gamma v^{s_2}(p)\right]\left[\bar{u}^{s_1}(k)\Gamma v^{s_2}(p)\right]^* = \mathrm{Tr}\left[\Gamma\left(\gamma^\alpha p_\alpha - m_1\right)\bar{\Gamma}\left(\gamma^\beta k_\beta + m_2\right)\right]$$

$$\sum_{s_1,s_2}\left[\bar{v}^{s_1}(k)\Gamma v^{s_2}(p)\right]\left[\bar{v}^{s_1}(k)\Gamma v^{s_2}(p)\right]^* = \mathrm{Tr}\left[\Gamma\left(\gamma^\alpha p_\alpha - m_1\right)\bar{\Gamma}\left(\gamma^\beta k_\beta - m_2\right)\right]$$

$$\sum_{s_1,s_2}\left[\bar{u}^{s_1}(k)\Gamma u^{s_2}(p)\right]\left[\bar{u}^{s_1}(k)\Gamma u^{s_2}(p)\right]^* = \mathrm{Tr}\left[\Gamma\left(\gamma^\alpha p_\alpha + m_1\right)\bar{\Gamma}\left(\gamma^\beta k_\beta + m_2\right)\right]$$

Inscribed Γ any one 4×4-Matrix, γ the Dirac matrices and a stroke of the Dirac adjunct $\bar{\Gamma} = \gamma^0\Gamma^\dagger\gamma^0$, m denote the masses of the respective particles/antiparticles.

Mathematical Background

The Dirac spinors can be divided into two independent spinors u for particles and disassemble for antiparticles. Each fulfills a completeness relation,

$$\sum_{\text{Spins}} u(p)\bar{u}(p) = \gamma^\mu p_\mu + m$$

$$\sum_{\text{Spins}} v(k)\bar{v}(k) = \gamma^\mu k_\mu - m.$$

Typical expressions such as for example come in the matrix element $\mathcal{M} \propto \bar{v}(k)\Gamma u(p)$ in front. The squared matrix element is therefore:

$$|\mathcal{M}|^2 \propto \left[\bar{v}(k)\Gamma u(p)\right]\left[\bar{v}(k)\Gamma u(p)\right]^* = \bar{v}(k)\Gamma u(p)\bar{u}(p)\bar{\Gamma}v(k)$$

When adding up over the spin indices, the completeness relation can first be applied in the middle pair of Dirac spinors. In the following, it is advisable not to suppress the spin and spacetime indices for reasons of comprehensibility, whereby s_1, s_2 about the spins, α, β about the four Dirac matrices and μ, ν, ρ, σ sum over the four components of the spinors:

$$\sum_{s_1,s_2}|\mathcal{M}|^2 \propto \sum_{s_1,s_2}\bar{v}_\mu^{s_1}(k)\Gamma^{\mu\nu}u_\nu^{s_2}(p)\bar{u}_\rho^{s_2}(p)\bar{\Gamma}^{\rho\sigma}v_\sigma^{s_1}(k) = \sum_{s_1}\bar{v}_\mu^{s_1}(k)\Gamma^{\mu\nu}\left(\gamma^\alpha p_\alpha + m_2\right)_{\nu\rho}\bar{\Gamma}^{\rho\sigma}v_\sigma^{s_1}(k)$$

In component notation, it is more obvious that the summation is over s_1 can be easily carried out since all objects are now commutating; it therefore applies,

$$\sum_{s_1,s_2}|\mathcal{M}|^2 \propto \Gamma^{\mu\nu}\left(\gamma^\alpha p_\alpha + m_2\right)_{\nu\rho}\bar{\Gamma}^{\rho\sigma}\left(\gamma^\beta k_\beta - m_1\right)_{\sigma\mu} = \mathrm{Tr}\left[\Gamma\left(\gamma^\alpha p_\alpha + m_2\right)\bar{\Gamma}\left(\gamma^\beta k_\beta - m_1\right)\right]$$

For the other three cases, the proof is analogous.

For Example and Electron-muon Scattering

Describe $p[p']$ the momentum of the electron going in and $k[k']$ that of the inbound muon is the matrix element of the electron-muon scattering $e^-\mu^- \to e^-\mu^-$ in the lowest order in quantum electrodynamics:

$$\mathcal{M} = \bar{u}(p)(ie\gamma^{\mu})u(p')\frac{ig_{\mu\nu}}{(p+k)^2}\bar{u}(k')(ie\gamma^{\nu})u(k)$$

If one averages over the spins of the incoming particles and adds up over the spins of the outgoing particles, the result of using Casimir's trick twice,

$$\frac{1}{4}\sum_{s_1,s_2,s_1',s_2'}|\mathcal{M}|^2 = \frac{1}{4}\frac{e^4}{(p+k)^4}\mathrm{Tr}\left[\gamma^{\mu}(\gamma^{\alpha}p_{\alpha}+m_e)\gamma^{\nu}(\gamma^{\beta}p'_{\beta}+m_e)\right]\mathrm{Tr}\left[\gamma_{\mu}(\gamma^{\alpha}k_{\alpha}+m_{\mu})\gamma_{\nu}(\gamma^{\beta}k'_{\beta}+m_{\mu})\right]$$

Quantum Chromodynamics 4

- **Quantum Chromodynamics Matter**
- **Lattice Quantum Chromodynamics**

Quantum chromodynamics (QCD) is a theory which deals with the strong interaction between glucons and quarks. QCD matter and lattice QCD are some of the methods that fall under its domain. The chapter closely examines these methods of quantum chromodynamics to provide an extensive understanding of the subject.

Quantum chromodynamics (QCD) is the component of the Standard Model that describes the strong interactions. QCD is the theory of quarks and gluons. Quarks carry a new charge, called color, that enables them to emit and absorb gluons. (This is the origin of the name chromodynamics, although the "color" of QCD should in no way be confused with the colors of light.) The quarks are also electrically charged and like electrons, are fermions, and carry a spin, or intrinsic angular momentum, of one-half in units of Planck's constant. The gluons are electrically neutral, and, like photons, are bosons of spin one. Together, the fields of quarks and gluons make up a nonabelian gauge theory.

In QCD, quarks interact by the exchange of gluons in much the same way that electrons interact by the exchange of the quanta of light, photons. Like photons, the gluons have no mass and travel at the speed of light. Unlike photons, however, the gluons carry the very color charge that produces them, so gluons can emit and absorb more gluons. The resulting strong force is thus more complicated to analyze than the electromagnetic force.

A convenient measure of the strength of the strong force is the QCD coupling $\alpha_s(Q)$ that controls the probability of a quark emitting a gluon, which produces forces between quarks. The QCD coupling depends on the momentum carried by the emitted gluon, denoted by Q. The strong coupling is large for very low-momentum gluons and decreases as the momentum increases, a variation known as asymptotic freedom. For the highest-momentum gluons that can be produced in modern accelerators, $\alpha_s(Q)$ is relatively small, about 0.1, but at momentum transfers characteristic of nuclear interactions, it gets to be quite large. Asymptotic freedom makes it easier to analyze processes over short times, which generally involve a few gluons, than over long times, which generally involve many.

Quarks come in six varieties, known as quark flavors. In the Standard Model, the six flavors of quarks, together with the six leptons (the electron, muon, tau, and their neutrinos), are truly elementary. The different flavors of quarks have different charges. Three quarks have electric charge

$+2e/3$: *the up* (u), charm (c) and top (t) quarks. Three quarks have charge $-e/3$ the down (d), strange (s) and bottom (b) quarks; -e is the charge of an electron. The masses of these quarks vary greatly, and of the six, only the u and d quarks, which are by far the lightest, appear to play a direct role in normal matter.

Hadrons

QCD binds quarks together into states that can be observed directly in the laboratory as hadrons, particles that feel the strong force. The best-known examples of hadrons are the nucleons, the proton and neutron, from which all atomic nuclei are formed. The idea of quarks arose to explain the regularities of hadron states, and their charges and spins could be readily explained (and even predicted) by simply combining the then known u, d, and s quarks. This is the quark model. Perhaps the most extraordinary feature of quarks in QCD is the confinement of quarks in hadrons. A free quark, one that is separated from a nucleon, would be readily detectable because its charge would *be* $-⅔$ *or* ⅓ that of the charge of an electron. No convincing evidence for such a particle has been found, and it is now believed that confinement is an unavoidable consequence of QCD.

There must be three quarks to make a proton or a neutron. The proton and neutron are different because the proton is a combination of two u quarks and one d quark and hence has a total charge of $2(2e/3) + (-e/3) = +e$. The neutron is made up of one u and two d quarks and hence has total charge $(2e/3) + 2(-e/3) = 0$.

In addition to nucleons, other combinations of three quarks have been observed, and all are known collectively as baryons. For example, from the u, d, and s quarks, it is possible to make ten distinct combinations, and all have been seen (notice that they all have charges that are integer multiples of e). In addition, baryons with c and b quarks have also been observed. All the baryons except for the proton and neutron decay quite rapidly because the weak interactions make all the quarks aside from the u and d unstable. The spins of the quarks that combine to form baryons may line up in parallel and antiparallel combinations, as long as the resulting state obeys the Pauli exclusion principle, which states that no two fermions may have the same set of quantum numbers. As a result, some of the baryons have total spin 1/2, and others spin 3/2, and all the baryons are fermions with half-integer total spin. With orbital angular momentum taken into account, even higher spins are possible, although these baryons are very unstable.

In addition to baryons, quarks can combine with their antiparticles, the antiquarks, to form mesons. Antiquarks are usually denoted by an overbar, such as ū for the antiquark of the $\hat{u}$. For example, the combinations $\hat{u}$ and $\overline{d}u$ can form the π^+ and π^+ the pions, of electric charge ^+e and ^-e, respectively. All the mesons are bosons with integer spins. Other exotic bosons, which are bound states of gluons without quarks, called glueballs, appear to be possible, although very unstable.

Color Charge and Gauge Theory

The concept of color was introduced to solve a problem in the quark model. The lowest-energy states in the quark model appeared to require that the three quarks in the proton be in identical states, which would violate the Pauli principle. With color, this conflict is avoided. For example, two u quarks can have the same energy and spin quantum numbers as long as their colors are different. Consistency with the Paul principle requires that all the quarks in any baryon have different colors: three quarks, three colors.

QCD is an example of a nonabelian gauge theory. In a nonabelian gauge theory, the concept of charge must be generalized from electromagnetism. An electric charge is just a number, such as e or $2e/3$, and the electric charge stays the same when a charged particle emits or absorbs a photon. A quark, however, actually has three separate charges that make up its color, which are sometimes labeled by the names of colors of light: a red (r) charge, a green (g) charge, and a blue (b) one. When a quark emits a gluon, its color charges change, and the kinds of gluons can be identified by combinations of the color "before" and the color "after," for example $r\bar{b}$, where the color with the overbar is the final color of the quark.

Correspondingly, when a quark with only color b absorbs an $\bar{r}b$ gluon, it changes into a quark with only color r. This way, the total of r, g, and b charges are conserved, and nine possibilities exist for the gluons. Of these nine, one combination, equal parts of $\bar{r}r$, $\bar{g}g$, and $\bar{b}b$, leaves all three of the colors the same and is absent, leaving eight gluons. The color charges in QCD have a surprising property: they can never be distinguished experimentally, and yet their number, three, can be measured. All hadrons have zero net r, b, and g colors; they are all colorless.

Evidence for Color, Quarks and Gluon

Experimental discoveries beginning in the late 1960s and early 1970s established firmly the reality of hadrons made up of quarks and gluons. As seen, quarks carry an electric charge, as do electrons. When an energetic electron collides with a target, the electron, which is blind to the strong force, nevertheless scatters from nucleons within the target by exchanging a photon. When the energy of the electron is high, the momentum p′ of the photon can be large, so large that the corresponding wavelength of light is much smaller than a nucleon in the target. This wavelength is given by $\lambda.=h/p$, where h is Planck's constant. The rules of quantum mechanics indicate that a photon cannot be absorbed by an object that is larger than its wavelength λ.

Nevertheless, the photons are absorbed at a rate almost independent of p, a phenomenon called scaling. When this happens, the nucleon generally breaks apart into high-energy fragments, a process called deep-inelastic scattering. The scaling of deep-inelastic scattering indicates that there are charged particles within nucleons that are much smaller than nucleons. These are the quarks. In addition, the distribution in angles of the scattered electrons depends on the spin of the charged particles. The electron-nucleon experiments show that all the charge in the nucleons is carried by spin-½ particles, exactly as suggested by the quark model.

In another set of experiments, electrons and their antiparticles, positrons, collide and annihilate into a virtual state that consists of a single photon. According to the rules of quantum mechanics, this photon then transforms itself into any pair of particle and antiparticle with nonzero electric charge, say, q. The probability for each species is proportional to , and the angular distribution at which they emerge depends on their spin. The quarks created this way are not observed directly because of confinement, but they each quickly evolve into jets of hadrons, which preserve their original directions. The total probability to produce hadrons is given simply by the probability to produce a single fermion of electric charge equal to 1, times the sum of the squares of the electric charges of all the quarks light enough to be produced at the available energy, times 3. The 3 stands for the three possible colors of each quark.

Many other high-energy experiments have confirmed the reality of quarks, gluons, and color. For example, a fraction of annihilation events include an extra jet from a gluon in addition to their quark and antiquark jets. In proton-proton scattering, jets also emerge from collisions, whose angular distributions exactly match predictions based on the elastic scattering of quarks through the exchange of gluons. Each of these predictions depends directly on the numbers of quark and gluon colors as well as their spins.

Physicists' understanding of the confinement and other low-energy properties of QCD is somewhat less complete but is still convincing. Qualitatively, it appears that the "empty" space—the vacuum—is not empty of QCD content. Analysis suggests that the vacuum acts as a sort of QCD superconductor, which repels the QCD lines of force between quarks, the QCD analogs of magnetic and electric fields between electrons. Within hadrons, this repulsion is absent. The larger the separation between isolated quarks, however, the larger the energy necessary to overcome the repulsion. Very quickly, it becomes easier to create enough pairs of quarks and antiquarks to hide all the lines of force within colorless hadrons than to lengthen them through the resistant vacuum.

Beginning in the 1970s, it became possible to simulate QCD on computers. Over time, more and more precise calculations have established that the energy of an isolated quark is essentially infinite and that the observed hadrons are, indeed, combinations of confined quarks. Experiments in which entire nuclei collide may create conditions in which quarks and gluons are temporarily freed, or deconfined. This would be a novel state of matter, called the quark-gluon plasma.

It is now understood that the Standard Model actually requires that there be three colors to avoid quantum inconsistencies, called anomalies, which would ruin it as a quantum theory. Nevertheless, physicists are far from a complete understanding of QCD. For example, quantum fluctuations in the QCD vacuum appear to have the ability to act differently on particles and antiparticles in a manner that is not seen in nature, a puzzle known as the strong CP problem. In this, and in how to reconcile fully the quark-gluon and baryon-meson descriptions of the strong interactions, we have much still to learn about quantum chromodynamics.

Every field theory of particle physics is based on certain symmetries of nature whose existence is deduced from observations. These can be:

- Local symmetries, that are the symmetries that act independently at each point in spacetime. Each such symmetry is the basis of a gauge theory and requires the introduction of its own gauge bosons.

- Global symmetries, which are symmetries whose operations must be simultaneously applied to all points of spacetime.

QCD is a non-abelian gauge theory (or Yang-Mills theory) of the SU(3) gauge group obtained by taking the color charge to define a local symmetry.

Since the strong interaction does not discriminate between different flavors of quark, QCD has approximate flavor symmetry, which is broken by the differing masses of the quarks.

There are additional global symmetries whose definitions require the notion of chirality, discrimination between left and right-handed. If the spin of a particle has a positive projection on its direction of motion then it is called left-handed; otherwise, it is right-handed. Chirality and handedness are not the same, but become approximately equivalent at high energies:

- Chiral symmetries involve independent transformations of these two types of particle.
- Vector symmetries (also called diagonal symmetries) mean the same transformation is applied on the two chiralities.
- Axial symmetries are those in which one transformation is applied on left-handed particles and the inverse on the right-handed particles.

Duality

As mentioned, *asymptotic freedom* means that at large energy – this corresponds also to *short distances* – there is practically no interaction between the particles. This is in contrast – more precisely one would say *dual*– to what one is used to, since usually one connects the absence of interactions with *large* distances. However, as already mentioned in the original paper of Franz Wegner, a solid state theorist who introduced 1971 simple gauge invariant lattice models, the high-temperature behaviour of the *original model*, e.g. the strong decay of correlations at large distances, corresponds to the low-temperature behaviour of the (usually ordered!) *dual model*, namely the asymptotic decay of non-trivial correlations, e.g. short-range deviations from almost perfect arrangements, for short distances.

Symmetry Groups

The color group SU(3) corresponds to the local symmetry whose gauging gives rise to QCD. The electric charge labels a representation of the local symmetry group U(1) which is gauged to give QED: this is an abelian group. If one considers a version of QCD with N_f flavors of massless quarks, then there is a global (chiral) flavor symmetry group $SU_L(N_f) \times SU_R(N_f) \times U_B(1) \times U_A(1)$ The chiral symmetry is spontaneously broken by the QCD vacuum to the vector $(L+R)$ $SU_V(N_f)$ with the formation of a chiral condensate. The vector symmetry, $U_B(1)$ corresponds to the baryon number of quarks and is an exact symmetry. The axial symmetry $U_A(1)$is exact in the classical theory, but broken in the quantum theory, an occurrence called an anomaly. Gluon field configurations called instantons are closely related to this anomaly.

There are two different types of SU(3) symmetry: there is the symmetry that acts on the different colors of quarks, and this is an exact gauge symmetry mediated by the gluons, and there is also a flavor symmetry which rotates different flavors of quarks to each other, or *flavor SU(3)*. Flavor SU(3) is an approximate symmetry of the vacuum of QCD, and is not a fundamental symmetry at all. It is an accidental consequence of the small mass of the three lightest quarks.

In the QCD vacuum there are vacuum condensates of all the quarks whose mass is less than the QCD scale. This includes the up and down quarks, and to a lesser extent the strange quark, but not any of the others. The vacuum is symmetric under SU(2) isospin rotations of up and down, and to a lesser extent under rotations of up, down and strange, or full flavor group SU(3), and the observed particles make isospin and SU(3) multiplets.

The approximate flavor symmetries do have associated gauge bosons, observed particles like the rho and the omega, but these particles are nothing like the gluons and they are not massless. They are emergent gauge bosons in an approximate string description of QCD.

Lagrangian

The dynamics of the quarks and gluons are controlled by the quantum chromodynamics Lagrangian. The gauge invariant QCD Lagrangian is

$$\mathcal{L}_{\text{QCD}} = \bar{\psi}_i \left(i(\gamma^\mu D_\mu)_{ij} - m\delta_{ij} \right) \psi_j - \frac{1}{4} G^a_{\mu\nu} G^{\mu\nu}_a$$

where $\psi_i(x)$ is the quark field, a dynamical function of spacetime, in the fundamental representation of the SU(3) gauge group, indexed by $i, j, \ldots; D_\mu$; is the gauge covariant derivative; the γ^μ are Dirac matrices connecting the spinor representation to the vector representation of the Lorentz group.

The symbol $G^a_{\mu\nu}$ represents the gauge invariant gluon field strength tensor, analogous to the electromagnetic field strength tensor, $F^{\mu\nu}$, in quantum electrodynamics. It is given by:

$$G^a_{\mu\nu} = \partial_\mu \mathcal{A}^a_\nu - \partial_\nu \mathcal{A}^a_\mu + g f^{abc} \mathcal{A}^b_\mu \mathcal{A}^c_\nu,$$

where $\mathcal{A}^a_\mu(x)$ are the gluon fields, dynamical functions of spacetime, in the adjoint representation of the SU(3) gauge group, indexed by $a, b, \ldots$ and f_{abc} are the structure constants of SU(3). Note that the rules to move-up or pull-down the *a*, *b*, or *c* indices are *trivial*, $(+, \ldots, +)$ so that $f^{abc} = f_{abc} = f^a{}_{bc}$ whereas for the μ or ν indices one has the non-trivial *relativistic* rules corresponding to the metric signature $(+ \; - \; - \; -)$.

The variables m and g correspond to the quark mass and coupling of the theory, respectively, which are subject to renormalization.

An important theoretical concept is the *Wilson loop* (named after Kenneth G. Wilson). In lattice QCD, the final term of the above Lagrangian is discretized via Wilson loops, and more generally the behavior of Wilson loops can distinguish confined and deconfined phases.

Fields

Quarks are massive spin-$\frac{1}{2}$ fermions which carry a color charge whose gauging is the content of QCD. Quarks are represented by Dirac fields in the fundamental representation 3 of the gauge group SU(3). They also carry electric charge (either $-\,^1/_3$ or $+^2/_3$) and participate in weak interactions as part of weak isospin doublets. They carry global quantum numbers including the baryon number, which is $^1/_3$ for each quark, hypercharge and one of the flavor quantum numbers.

Gluons are spin-1 bosons which also carry color charges, since they lie in the adjoint representation 8 of SU(3). They have no electric charge, do not participate in the weak interactions, and have no flavor. They lie in the singlet representation 1 of all these symmetry groups.

Every quark has its own antiquark. The charge of each antiquark is exactly the opposite of the corresponding quark.

Dynamics

According to the rules of quantum field theory, and the associated Feynman diagrams, the above theory gives rise to three basic interactions: a quark may emit (or absorb) a gluon, a gluon may emit (or absorb) a gluon, and two gluons may directly interact. This contrasts with QED, in which only the first kind of interaction occurs, since photons have no charge. Diagrams involving Faddeev–Popov ghosts must be considered too (except in the unitarity gauge).

Area Law and Confinement

Detailed computations with the above-mentioned Lagrangian show that the effective potential between a quark and its anti-quark in a meson contains a term that increases in proportion to the distance between the quark and anti-quark ($(\propto r)$), which represents some kind of "stiffness" of the interaction between the particle and its anti-particle at large distances, similar to the entropic elasticity of a rubber band. This leads to *confinement* of the quarks to the interior of hadrons, i.e. mesons and nucleons, with typical radii R_c, corresponding to former "Bag models" of the hadrons The order of magnitude of the "bag radius" is 1 fm $(=10^{-15}\,m)$ Moreover, the above-mentioned stiffness is quantitatively related to the so-called "area law" behaviour of the expectation value of the Wilson loop product P_W of the ordered coupling constants around a closed loop W; i.e. $\langle P_W \rangle$ is proportional to the *area* enclosed by the loop. For this behaviour the non-abelian behaviour of the gauge group is essential.

Methods

Further analysis of the content of the theory is complicated. Various techniques have been developed to work with QCD.

Perturbative QCD

This approach is based on asymptotic freedom, which allows perturbation theory to be used accurately in experiments performed at very high energies. Although limited in scope, this approach has resulted in the most precise tests of QCD to date.

Lattice QCD

Among non-perturbative approaches to QCD, the most well established one is lattice QCD. This approach uses a discrete set of spacetime points (called the lattice) to reduce the analytically intractable path integrals of the continuum theory to a very difficult numerical computation which is then carried out on supercomputers like the QCDOC which was constructed for precisely this purpose. While it is a slow and resource-intensive approach, it has wide applicability, giving insight into parts of the theory inaccessible by other means, in particular into the explicit forces acting between quarks and antiquarks in a meson. However, the numerical sign problem makes it difficult to use lattice methods to study QCD at high density and low temperature (e.g. nuclear matter or the interior of neutron stars).

$1/_N$ Expansion

A well-known approximation scheme, the $^1/_N$ expansion, starts from the idea that the number of colors is infinite, and makes a series of corrections to account for the fact that it is not. Until now, it has been the source of qualitative insight rather than a method for quantitative predictions. Modern variants include the AdS/CFT approach.

Effective Theories

For specific problems effective theories may be written down which give qualitatively correct results in certain limits. In the best of cases, these may then be obtained as systematic expansions in some parameter of the QCD Lagrangian. One such effective field theory is chiral perturbation theory or ChiPT, which is the QCD effective theory at low energies. More precisely, it is a low energy expansion based on the spontaneous chiral symmetry breaking of QCD, which is an exact symmetry when quark masses are equal to zero, but for the u, d and s quark, which have small mass, it is still a good approximate symmetry. Depending on the number of quarks which are treated as light, one uses either SU(2) ChiPT or SU(3) ChiPT. Other effective theories are heavy quark effective theory (which expands around heavy quark mass near infinity), and soft-collinear effective theory (which expands around large ratios of energy scales). In addition to effective theories, models like the Nambu–Jona-Lasinio model and the chiral model are often used when discussing general features.

QCD Sum Rules

Based on an Operator product expansion one can derive sets of relations that connect different observables with each other.

Nambu–Jona-Lasinio Model

In one of his recent works, Kei-Ichi Kondo derived as a low-energy limit of QCD, a theory linked to the Nambu–Jona-Lasinio model since it is basically a particular non-local version of the Polyakov–Nambu–Jona-Lasinio model. The later being in its local version, nothing but the Nambu–Jona-Lasinio model in which one has included the Polyakov loop effect, in order to describe a 'certain confinement'.

The Nambu–Jona-Lasinio model in itself is, among many other things, used because it is a 'relatively simple' model of chiral symmetry breaking, phenomenon present up to certain conditions (Chiral limit i.e. massless fermions) in QCD itself. In this model, however, there is no confinement. In particular, the energy of an isolated quark in the physical vacuum turns out well defined and finite.

Experimental Tests

The notion of quark flavors was prompted by the necessity of explaining the properties of hadrons during the development of the quark model. The notion of color was necessitated by the puzzle of the Δ^{++}.

The first evidence for quarks as real constituent elements of hadrons was obtained in deep inelastic scattering experiments at SLAC. The first evidence for gluons came in three jet events at PETRA.

Several good quantitative tests of perturbative QCD exist:

- The running of the QCD coupling as deduced from many observations.
- Scaling violation in polarized and unpolarized deep inelastic scattering.
- Vector boson production at colliders (this includes the Drell-Yan process).
- Direct photons produced in hadronic collisions.
- Jet cross sections in colliders.
- Event shape observables at the LEP.
- Heavy-quark production in colliders.

Quantitative tests of non-perturbative QCD are fewer, because the predictions are harder to make. The best is probably the running of the QCD coupling as probed through lattice computations of heavy-quarkonium spectra. There is a recent claim about the mass of the heavy meson B_c. Other non-perturbative tests are currently at the level of 5% at best. Continuing work on masses and form factors of hadrons and their weak matrix elements are promising candidates for future quantitative tests. The whole subject of quark matter and the quark–gluon plasma is a non-perturbative test bed for QCD which still remains to be properly exploited.

One qualitative prediction of QCD is that there exist composite particles made solely of gluons called glueballs that have not yet been definitively observed experimentally. A definitive observation of a glueball with the properties predicted by QCD would strongly confirm the theory. In principle, if glueballs could be definitively ruled out, this would be a serious experimental blow to QCD. But, as of 2013, scientists are unable to confirm or deny the existence of glueballs definitively, despite the fact that particle accelerators have sufficient energy to generate them.

Cross-relations to Solid State Physics

There are unexpected cross-relations to solid state physics. For example, the notion of gauge invariance forms the basis of the well-known Mattis spin glasses, which are systems with the usual spin degrees of freedom $s_i = \pm 1$ for $i = 1,\ldots,N,$, with the special fixed "random" couplings $J_{i,k} = \epsilon_i J_0 \epsilon_k$. Here the ε_i and ε_k quantities can independently and "randomly" take the values ±1, which corresponds to a most-simple gauge transformation $(s_i \to s_i \cdot \epsilon_i \quad J_{i,k} \to \epsilon_i J_{i,k} \epsilon_k \quad s_k \to s_k \cdot \epsilon_k)$. This means that thermodynamic expectation values of measurable quantities, e.g. of the energy $\mathcal{H} := -\sum s_i J_{i,k} s_k$, are invariant.

However, here the *coupling degrees of freedom* $J_{i,k}$, which in the QCD correspond to the *gluons*, are "frozen" to fixed values (quenching). In contrast, in the QCD they "fluctuate" (annealing), and through the large number of gauge degrees of freedom the entropy plays an important role

For positive J_0 the thermodynamics of the Mattis spin glass corresponds in fact simply to a "ferromagnet in disguise", just because these systems have no "frustration" at all. This term is a basic measure in spin glass theory. Quantitatively it is identical with the loop product $P_W := J_{i,k} J_{k,l} \ldots J_{n,m} J_{m,i}$ along a closed loop W. However, for a Mattis spin glass – in contrast to "genuine" spin glasses – the quantity P_W never becomes negative.

The basic notion "frustration" of the spin-glass is actually similar to the Wilson loop quantity of the QCD. The only difference is again that in the QCD one is dealing with SU(3) matrices, and that one is dealing with a "fluctuating" quantity. Energetically, perfect absence of frustration should be non-favorable and atypical for a spin glass, which means that one should add the loop product to the Hamiltonian, by some kind of term representing a "punishment". In the QCD the Wilson loop is essential for the Lagrangian rightaway.

The relation between the QCD and "disordered magnetic systems" (the spin glasses belong to them) were additionally stressed in a paper by Fradkin, Huberman and Shenker, which also stresses the notion of duality.

A further analogy consists in the already mentioned similarity to polymer physics, where, analogously to Wilson Loops, so-called "entangled nets" appear, which are important for the formation of the entropy-elasticity (force proportional to the length) of a rubber band. The non-abelian character of the SU(3) corresponds thereby to the non-trivial "chemical links", which glue different loop segments together, and "asymptotic freedom" means in the polymer analogy simply the fact that in the short-wave limit, i.e. for $0 \leftarrow \lambda_w \ll R_c$ (where R_c is a characteristic correlation length for the glued loops, corresponding to the above-mentioned "bag radius", while λ_w is the wavelength of an excitation) any non-trivial correlation vanishes totally, as if the system had crystallized.

There is also a correspondence between confinement in QCD – the fact that the color field is only different from zero in the interior of hadrons – and the behaviour of the usual magnetic field in the theory of type-II superconductors: there the magnetism is confined to the interior of the Abrikosov flux-line lattice, i.e., the London penetration depth λ of that theory is analogous to the confinement radius R_c of quantum chromodynamics. Mathematically, this correspondendence is supported by the second term, $\propto gG^a_\mu \bar{\psi}_i \gamma^\mu T^a_{ij} \psi_j$, on the r.h.s. of the Lagrangian.

Quantum Chromodynamics Matter

Quark matter or QCD matter (quantum chromodynamic) refers to any of a number of phases of matter whose degrees of freedom include quarks and gluons. These phases occurs at extremely high temperatures and densities, and some of them are still only theoretical as they require conditions so extreme that they can not be produced in any laboratory, especially not as equilibrium conditions. Under these extreme conditions, the familiar structure of matter, where the basic constituents are nuclei (consisting of nucleons which are bound states of quarks) and electrons, is disrupted. In quark matter it is more appropriate to treat the quarks themselves as the basic degrees of freedom.

In the standard model of particle physics, the strong force is described by the theory of QCD. At ordinary temperatures or densities this force just confines the quarks into composite particles (hadrons) of size around 10^{-15} m = 1 femtometer = 1 fm (corresponding to the QCD energy scale $\Lambda_{QCD} \approx 200$ MeV) and its effects are not noticeable at longer distances. However, when the temperature reaches the QCD energy scale (T of order 10^{12} kelvins) or the density rises to the point where the average inter-quark separation is less than 1 fm (quark chemical potential μ around

400 MeV), the hadrons are melted into their constituent quarks, and the strong interaction becomes the dominant feature of the physics. Such phases are called quark matter or QCD matter.

The strength of the color force makes the properties of quark matter unlike gas or plasma, instead leading to a state of matter more reminiscent of a liquid. At high densities, quark matter is a Fermi liquid, but is predicted to exhibit color superconductivity at high densities and temperatures below 10^{12} K.

Occurrence

Natural Occurrence

According to the Big Bang theory, in the early universe at high temperatures when the universe was only a few tens of microseconds old, the phase of matter took the form of a hot phase of quark matter called the quark–gluon plasma (QGP).

- Compact stars (neutron stars): A neutron star is much cooler than 10^{12} K, but it has been compressed by the supernova creating it to such high densities, that it is reasonable to surmise that quark matter may exist in the core. Compact stars composed mostly or entirely of quark matter are called quark stars or strange stars, yet at this time no star with properties expected of these objects has been observed.
- Strangelets: These are theoretically postulated (but as yet unobserved) lumps of strange matter comprising nearly equal amounts of up, down and strange quarks. Strangelets are supposed to be present in the galactic flux of high energy particles and should therefore theoretically be detectable in cosmic rays here on Earth, but no strangelet has been detected with certainty.
- Cosmic ray impacts: Cosmic rays comprise a lot of different particles, including highly accelerated atomic nuclei, particularly that of iron. Laboratory experiments suggests that the inevitable interaction with heavy noble gas nuclei in the upper atmosphere would lead to quark–gluon plasma formation.
- Small pieces of quark matter with baryon number over about 300 may be more stable than nuclear matter, and form what appears to be heavy nuclei, with higher atomic numbers than known elements. This form of atomic matter could possibly form a continent of stability.

Laboratory Experiments

Even though quark-gluon plasma can only occur under quite extreme conditions of temperature and/or pressure, it is being actively studied at particle colliders, such as the Large Hadron Collider LHC at CERN and the Relativistic Heavy Ion Collider RHIC at Brookhaven National Laboratory. In these collisions, the plasma only occurs for a very short time before it spontaneously disintegrates, because the extreme conditions during the collision process cannot be upheld. The plasma's physical characteristics are studied by detecting the debris emanating from the collision region with large particle detectors.

Heavy-ion collisions at very high energies can produce small short-lived regions of space whose energy density is comparable to that of the 20-micro-second-old universe. This has been achieved by colliding heavy nuclei such as lead nuclei at high speeds, and a first time claim of formation of quark–gluon plasma came from the SPS accelerator at CERN in February 2000. This work has been continued at more powerful accelerators, such as RHIC in the US, and as of 2010 at the European LHC at CERN located in the border area of Switzerland and France. There is good evidence that the quark–gluon plasma has also been produced at RHIC.

Thermodynamics

The context for understanding the thermodynamics of quark matter is the standard model of particle physics, which contains six different flavors of quarks, as well as leptons like electrons and neutrinos. These interact via the strong interaction, electromagnetism, and also the weak interaction which allows one flavor of quark to turn into another. Electromagnetic interactions occur between particles that carry electrical charge; strong interactions occur between particles that carry color charge.

The correct thermodynamic treatment of quark matter depends on the physical context. For large quantities that exist for long periods of time (the "thermodynamic limit"), we must take into account the fact that the only conserved charges in the standard model are quark number (equivalent to baryon number), electric charge, the eight color charges, and lepton number. Each of these can have an associated chemical potential. However, large volumes of matter must be electrically and color-neutral, which determines the electric and color charge chemical potentials. This leaves a three-dimensional phase space, parameterized by quark chemical potential, lepton chemical potential, and temperature.

In compact stars quark matter would occupy cubic kilometers and exist for millions of years, so the thermodynamic limit is appropriate. However, the neutrinos escape, violating lepton number, so the phase space for quark matter in compact stars only has two dimensions, temperature (T) and quark number chemical potential μ. A strangelet is not in the thermodynamic limit of large volume, so it is like an exotic nucleus: it may carry electric charge.

A heavy-ion collision is in neither the thermodynamic limit of large volumes nor long times. Putting aside questions of whether it is sufficiently equilibrated for thermodynamics to be applicable, there is certainly not enough time for weak interactions to occur, so flavor is conserved, and there are independent chemical potentials for all six quark flavors. The initial conditions (the impact parameter of the collision, the number of up and down quarks in the colliding nuclei, and the fact that they contain no quarks of other flavors) determine the chemical potentials.

Phase Diagram

The phase diagram of quark matter is not well known, either experimentally or theoretically. A commonly conjectured form of the phase diagram is shown in the figure. It is applicable to matter in a compact star, where the only relevant thermodynamic potentials are quark chemical potential μ and temperature T. For guidance it also shows the typical values of μ and T in heavy-ion collisions and in the early universe. For readers who are not familiar with the concept of a chemical potential, it is helpful to think of μ as a measure of the imbalance between quarks and antiquarks in the system. Higher μ means a stronger bias favoring quarks over antiquarks. At low temperatures there are no antiquarks, and then higher μ generally means a higher density of quarks.

Ordinary atomic matter as we know it is really a mixed phase, droplets of nuclear matter (nuclei) surrounded by vacuum, which exists at the low-temperature phase boundary between vacuum and nuclear matter, at $\mu = 310$ MeV and T close to zero. If we increase the quark density (i.e. increase μ) keeping the temperature low, we move into a phase of more and more compressed nuclear matter. Following this path corresponds to burrowing more and more deeply into a neutron star. Eventually, at an unknown critical value of μ, there is a transition to quark matter. At ultra-high densities we expect to find the color-flavor-locked (CFL) phase of color-superconducting quark matter. At intermediate densities we expect some other phases (labelled "non-CFL quark liquid" in the figure) whose nature is presently unknown,. They might be other forms of color-superconducting quark matter, or something different.

Now, imagine starting at the bottom left corner of the phase diagram, in the vacuum where $\mu = T = 0$. If we heat up the system without introducing any preference for quarks over antiquarks, this corresponds to moving vertically upwards along the T axis. At first, quarks are still confined and we create a gas of hadrons (pions, mostly). Then around $T = 150$ MeV there is a crossover to the quark gluon plasma: thermal fluctuations break up the pions, and we find a gas of quarks, antiquarks, and gluons, as well as lighter particles such as photons, electrons, positrons, etc. Following this path corresponds to travelling far back in time (so to say), to the state of the universe shortly after the big bang (where there was a very tiny preference for quarks over antiquarks).

The line that rises up from the nuclear/quark matter transition and then bends back towards the T axis, with its end marked by a star, is the conjectured boundary between confined and unconfined phases. Until recently it was also believed to be a boundary between phases where chiral symmetry is broken (low temperature and density) and phases where it is unbroken (high temperature and density). It is now known that the CFL phase exhibits chiral symmetry breaking, and other quark matter phases may also break chiral symmetry, so it is not clear whether this is really a chiral transition line. The line ends at the "chiral critical point", marked by a star in this figure, which is a special temperature and density at which striking physical phenomena, analogous to critical opalescence, are expected.

For a complete description of phase diagram it is required that one must have complete understanding of dense, strongly interacting hadronic matter and strongly interacting quark matter from some underlying theory e.g. quantum chromodynamics (QCD). However, because such a description requires the proper understanding of QCD in its non-perturbative regime, which is still far from being completely understood, any theoretical advance remains very challenging.

Theoretical Challenges: Calculation Techniques

The phase structure of quark matter remains mostly conjectural because it is difficult to perform calculations predicting the properties of quark matter. The reason is that QCD, the theory describing the dominant interaction between quarks, is strongly coupled at the densities and temperatures of greatest physical interest, and hence it is very hard to obtain any predictions from it. Here are brief descriptions of some of the standard approaches.

Lattice Gauge Theory

The only first-principles calculational tool currently available is lattice QCD, i.e. brute-force computer calculations. Because of a technical obstacle known as the fermion sign problem, this method can only be used at low density and high temperature ($\mu < T$), and it predicts that the crossover to the quark–gluon plasma will occur around $T = 150$ MeV However, it cannot be used to investigate the interesting color-superconducting phase structure at high density and low temperature.

Weak Coupling Theory

Because QCD is asymptotically free it becomes weakly coupled at unrealistically high densities, and diagrammatic methods can be used. Such methods show that the CFL phase occurs at very high density. At high temperatures, however, diagrammatic methods are still not under full control.

Models

To obtain a rough idea of what phases might occur, one can use a model that has some of the same properties as QCD, but is easier to manipulate. Many physicists use Nambu-Jona-Lasinio models, which contain no gluons, and replace the strong interaction with a four-fermion interaction. Mean-field methods are commonly used to analyse the phases. Another approach is the bag model, in which the effects of confinement are simulated by an additive energy density that penalizes unconfined quark matter.

Effective Theories

Many physicists simply give up on a microscopic approach, and make informed guesses of the expected phases (perhaps based on NJL model results). For each phase, they then write down an effective theory for the low-energy excitations, in terms of a small number of parameters, and use it to make predictions that could allow those parameters to be fixed by experimental observations.

Other Approaches

There are other methods that are sometimes used to shed light on QCD, but for various reasons have not yet yielded useful results in studying quark matter.

1/N Expansion

Treat the number of colors N, which is actually 3, as a large number, and expand in powers of $1/N$. It turns out that at high density the higher-order corrections are large, and the expansion gives misleading results.

Supersymmetry

Adding scalar quarks (squarks) and fermionic gluons (gluinos) to the theory makes it more tractable, but the thermodynamics of quark matter depends crucially on the fact that only fermions can carry quark number, and on the number of degrees of freedom in general.

Experimental Challenges

Experimentally, it is hard to map the phase diagram of quark matter because it has been rather difficult to learn how to tune to high enough temperatures and density in the laboratory experiment using collisions of relativistic heavy ions as experimental tools. However, these collisions ultimately will provide information about the crossover from hadronic matter to QGP. It has been suggested that the observations of compact stars may also constrain the information about the high-density low-temperature region. Models of the cooling, spin-down, and precession of these stars offer information about the relevant properties of their interior. As observations become more precise, physicists hope to learn more.

One of the natural subjects for future research is the search for the exact location of the chiral critical point. Some ambitious lattice QCD calculations may have found evidence for it, and future calculations will clarify the situation. Heavy-ion collisions might be able to measure its position experimentally, but this will require scanning across a range of values of μ and T.

Lattice Quantum Chromodynamics

Lattice QCD is a well-established non-perturbative approach to solving the quantum chromodynamics (QCD) theory of quarks and gluons. It is a lattice gauge theory formulated on a grid or lattice of points in space and time. When the size of the lattice is taken infinitely large and its sites infinitesimally close to each other, the continuum QCD is recovered.

Analytic or perturbative solutions in low-energy QCD are hard or impossible to obtain due to the highly nonlinear nature of the strong force and the large coupling constant at low energies. This formulation of QCD in discrete rather than continuous spacetime naturally introduces a momentum cut-off at the order $1/a$, where a is the lattice spacing, which regularizes the theory. As a result, lattice QCD is mathematically well-defined. Most importantly, lattice QCD provides a framework for investigation of non-perturbative phenomena such as confinement and quark–gluon plasma formation, which are intractable by means of analytic field theories.

In lattice QCD, fields representing quarks are defined at lattice sites (which leads to fermion doubling), while the gluon fields are defined on the links connecting neighboring sites. This approximation approaches continuum QCD as the spacing between lattice sites is reduced to zero. Because the computational cost of numerical simulations can increase dramatically as the lattice spacing decreases, results are often extrapolated to $a = 0$ by repeated calculations at different lattice spacings a that are large enough to be tractable.

Numerical lattice QCD calculations using Monte Carlo methods can be extremely computationally intensive, requiring the use of the largest available supercomputers. To reduce the computational burden, the so-called quenched approximation can be used, in which the quark fields are treated as non-dynamic "frozen" variables. While this was common in early lattice QCD calculations, "dynamical" fermions are now standard. These simulations typically utilize algorithms based upon molecular dynamics or microcanonical ensemble algorithms.

At present, lattice QCD is primarily applicable at low densities where the numerical sign problem does not interfere with calculations. Lattice QCD predicts that confined quarks will become released to quark-gluon plasma around energies of 150 MeV. Monte Carlo methods are free from the sign problem when applied to the case of QCD with gauge group SU(2) (QC_2D).

Lattice QCD has already made successful contact with many experiments. For example, the mass of the proton has been determined theoretically with an error of less than 2 percent.

Lattice QCD has also been used as a benchmark for high-performance computing, an approach originally developed in the context of the IBM Blue Gene supercomputer.

Techniques

Monte-Carlo Simulations

Monte-Carlo is a method to pseudo-randomly sample a large space of variables. The importance sampling technique used to select the gauge configurations in the Monte-Carlo simulation imposes the use of Euclidean time, by a Wick rotation of spacetime.

In lattice Monte-Carlo simulations the aim is to calculate correlation functions. This is done by explicitly calculating the action, using field configurations which are chosen according to the distribution function, which depends on the action and the fields. Usually one starts with the gauge bosons part and gauge-fermion interaction part of the action to calculate the gauge configurations, and then uses the simulated gauge configurations to calculate hadronic propagators and correlation functions.

Fermions on the Lattice

Lattice QCD is a way to solve the theory exactly from first principles, without any assumptions, to the desired precision. However, in practice the calculation power is limited, which requires a smart use of the available resources. One needs to choose an action which gives the best physical description of the system, with minimum errors, using the available computational power. The limited computer resources force one to use approximate physical constants which are different to their true physical values:

- The lattice discretization means approximating continuous and infinite space-time by a finite lattice spacing and size. The smaller the lattice, and the bigger the gap between nodes, the bigger the error. Limited resources commonly force the use of smaller physical lattices and larger lattice spacing than wanted, leading to larger errors than wanted.
- The quark masses are also approximated. Quark masses are larger than experimentally measured. These have been steadily approaching their physical values, and within the past few years a few collaborations have used nearly physical values to extrapolate down to physical values.

In order to compensate for the errors one improves the lattice action in various ways, to minimize mainly finite spacing errors.

Lattice Perturbation Theory

In lattice perturbation theory the scattering matrix is expanded in powers of the lattice spacing, a. The results are used primarily to renormalize Lattice QCD Monte-Carlo calculations. In perturbative calculations both the operators of the action and the propagators are calculated on the lattice and expanded in powers of a. When renormalizing a calculation, the coefficients of the expansion need to be matched with a common continuum scheme, such as the MS-bar scheme, otherwise the results cannot be compared. The expansion has to be carried out to the same order in the continuum scheme and the lattice one.

The lattice regularization was initially introduced by Wilson as a framework for studying strongly coupled theories non-perturbatively. However, it was found to be a regularization suitable also for perturbative calculations. Perturbation theory involves an expansion in the coupling constant, and is well-justified in high-energy QCD where the coupling constant is small, while it fails completely when the coupling is large and higher order corrections are larger than lower orders in the perturbative series. In this region non-perturbative methods, such as Monte-Carlo sampling of the correlation function, are necessary.

Lattice perturbation theory can also provide results for condensed matter theory. One can use the lattice to represent the real atomic crystal. In this case the lattice spacing is a real physical value, and not an artifact of the calculation which has to be removed, and a quantum field theory can be formulated and solved on the physical lattice.

Quantum Computing

In 2005 researchers of the National Institute of Informatics reformulated the $U(1), SU(2)$ *and* $SU(3)$ lattice gauge theories into a form that can be simulated using "spin qubit manipulations" on a universal quantum computer.

Limitations

The method suffers from a few limitations:

- Currently there is no formulation of lattice QCD that allows us to simulate the real-time dynamics of a quark-gluon system such as quark-gluon plasma.
- It is computationally intensive, with the bottleneck not being flops but the bandwidth of memory access.

References

- Quantum-chromodynamics, science-and-technology-physics: encyclopedia.com, Retrieved 05 February, 2019
- Kei-Ichi Kondo (2010). "Toward a first-principle derivation of confinement and chiral-symmetry-breaking crossover transitions in QCD". Physical Review D. 82 (6): 065024. arXiv:1005.0314. Bibcode:2010PhRvD..82f5024K. doi:10.1103/PhysRevD.82.065024
- Greiner, Walter; Schramm, Stefan; Stein, Eckart (2007). Quantum Chromodynamics. Berlin Heidelberg: Springer. ISBN 978-3-540-48535-3
- Holdom, Bob; Ren, Jing; Zhang, Chen (31 May 2018). "Quark Matter May Not Be Strange". Physical Review Letters. 120 (22): 222001. arXiv:1707.06610. Bibcode:2018PhRvL.120v2001H. doi:10.1103/PhysRevLett.120.222001
- P. Petreczky (2012). "Lattice QCD at non-zero temperature". J. Phys. G. 39 (9): 093002. arXiv:1203.5320. Bibcode:2012JPhG...39i3002P. doi:10.1088/0954-3899/39/9/093002

Strong Interactions 5

- **Nuclear Force**
- **Nucleons**
- **Mesons**
- **Isospin**
- **Hypercharge**
- **Baryons**
- **Gluons**

Strong interaction is referred to the mechanism that causes a strong nuclear force. The strong interactions of nucleons, isospin, gluons, baryons, etc. are studied within this subject. The topics elaborated in this chapter will help in gaining a better perspective about strong interactions associated with particle physics.

Strong force or Strong interaction is a fundamental interaction of nature that acts between subatomic particles of matter. The strong force binds quarks together in clusters to make more-familiar subatomic particles, such as protons and neutrons. It also holds together the atomic nucleus and underlies interactions between all particles containing quarks. The strong force originates in a property known as colour. This property, which has no connection with colour in the visual sense of the word, is somewhat analogous to electric charge. Just as electric charge is the source of electromagnetism, or the electromagnetic force, so colour is the source of the strong force. Particles without colour, such as electrons and other leptons, do not "feel" the strong force; particles with colour, principally the quarks, do "feel" the strong force. Quantum chromodynamics, the quantum field theory describing strong interactions, takes its name from this central property of colour.

Protons and neutrons are examples of baryons, a class of particles that contain three quarks, each with one of three possible values of colour (red, blue, and green). Quarks may also combine with antiquarks (their antiparticles, which have opposite colour) to form mesons, such as pi mesons and K mesons. Baryons and mesons all have a net colour of zero, and it seems that the strong force allows only combinations with zero colour to exist. Attempts to knock out individual quarks, in high-energy particle collisions, for example, result only in the creation of new "colourless" particles, mainly mesons.

In strong interactions the quarks exchange gluons, the carriers of the strong force. Gluons, like photons (the messenger particles of the electromagnetic force), are massless particles with a whole unit of intrinsic spin. However, unlike photons, which are not electrically charged and therefore do not feel the electromagnetic force, gluons carry colour, which means that they do feel the strong force and can interact among themselves. One result of this difference is that, within its short range (about 10^{-15} metre, roughly the diameter of a proton or a neutron), the strong force appears to become stronger with distance, unlike the other forces.

As the distance between two quarks increases, the force between them increases rather as the tension does in a piece of elastic as its two ends are pulled apart. Eventually the elastic will break, yielding two pieces. Something similar happens with quarks, for with sufficient energy it is not one quark but a quark-antiquark pair that is "pulled" from a cluster. Thus, quarks appear always to be locked inside the observable mesons and baryons, a phenomenon known as confinement. At distances comparable to the diameter of a proton, the strong interaction between quarks is about 100 times greater than the electromagnetic interaction. At smaller distances, however, the strong force between quarks becomes weaker, and the quarks begin to behave like independent particles, an effect known as asymptotic freedom.

Nuclear Force

The nuclear force (or nucleon–nucleon interaction or residual strong force) is a force that acts between the protons and neutrons of atoms. Neutrons and protons, both nucleons, are affected by the nuclear force almost identically. Since protons have charge +1 *e*, they experience an electric force that tends to push them apart, but at short range the attractive nuclear force is strong enough to overcome the electromagnetic force. The nuclear force binds nucleons into atomic nuclei.

The nuclear force is powerfully attractive between nucleons at distances of about 1 femtometre (fm, or 1.0×10^{-15} metres), but it rapidly decreases to insignificance at distances beyond about 2.5 fm. At distances less than 0.7 fm, the nuclear force becomes repulsive. This repulsive component is responsible for the physical size of nuclei, since the nucleons can come no closer than the force allows. By comparison, the size of an atom, measured in angstroms($A1.0 \times 10^{-10}$)is five orders of magnitude larger. The nuclear force is not simple, however, since it depends on the nucleon spins, has a tensor component, and may depend on the relative momentum of the nucleons. The strong nuclear force is one of the fundamental forces of nature.

The nuclear force plays an essential role in storing energy that is used in nuclear power and nuclear weapons. Work (energy) is required to bring charged protons together against their electric repulsion. This energy is stored when the protons and neutrons are bound together by the nuclear force to form a nucleus. The mass of a nucleus is less than the sum total of the individual masses of the protons and neutrons. The difference in masses is known as the mass defect, which can be expressed as an energy equivalent. Energy is released when a heavy nucleus breaks apart into two or more lighter nuclei. This energy is the electromagnetic potential energy that is released when the nuclear force no longer holds the charged nuclear fragments together.

A quantitative description of the nuclear force relies on equations that are partly empirical. These equations model the internucleon potential energies, or potentials. (Generally, forces within a system of particles can be more simply modeled by describing the system's potential energy; the negative gradient of a potential is equal to the vector force.) The constants for the equations are phenomenological, that is, determined by fitting the equations to experimental data. The internucleon potentials attempt to describe the properties of nucleon–nucleon interaction. Once determined, any given potential can be used in, e.g., the Schrödinger equation to determine the quantum mechanical properties of the nucleon system.

The discovery of the neutron in 1932 revealed that atomic nuclei were made of protons and neutrons, held together by an attractive force. By 1935 the nuclear force was conceived to be transmitted by particles called mesons. This theoretical development included a description of the Yukawa potential, an early example of a nuclear potential. Mesons, predicted by theory, were discovered experimentally in 1947. By the 1970s, the quark model had been developed, by which the mesons and nucleons were viewed as composed of quarks and gluons. By this new model, the nuclear force, resulting from the exchange of mesons between neighboring nucleons, is a residual effect of the strong force.

While the nuclear force is usually associated with nucleons, more generally this force is felt between hadrons, or particles composed of quarks. At small separations between nucleons (less than ~ 0.7 fm between their centers, depending upon spin alignment) the force becomes repulsive, which keeps the nucleons at a certain average separation, even if they are of different types. This repulsion arises from the Pauli exclusion force for identical nucleons (such as two neutrons or two protons). A Pauli exclusion force also occurs between quarks of the same type within nucleons, when the nucleons are different (a proton and a neutron, for example).

Field Strength

At distances larger than 0.7 fm the force becomes attractive between spin-aligned nucleons, becoming maximal at a center–center distance of about 0.9 fm. Beyond this distance the force drops exponentially, until beyond about 2.0 fm separation, the force is negligible. Nucleons have a radius of about 0.8 fm.

At short distances (less than 1.7 fm or so), the attractive nuclear force is stronger than the repulsive Coulomb force between protons; it thus overcomes the repulsion of protons within the nucleus. However, the Coulomb force between protons has a much greater range as it varies as the inverse square of the charge separation, and Coulomb repulsion thus becomes the only significant force between protons when their separation exceeds about 2 to 2.5 fm.

The nuclear force has a spin-dependent component. The force is stronger for particles with their spins aligned than for those with their spins anti-aligned. If two particles are the same, such as two neutrons or two protons, the force is not enough to bind the particles, since the spin vectors of two particles of the same type must point in opposite directions when the particles are near each other and are (save for spin) in the same quantum state. This requirement for fermions stems from the Pauli exclusion principle. For fermion particles of different types, such as a proton and neutron, particles may be close to each other and have aligned spins without violating the Pauli exclusion principle, and the nuclear force may bind them (in this case, into a deuteron), since the nuclear force is much stronger for spin-aligned particles. But if the particles' spins are anti-aligned the nuclear force is too weak to bind them, even if they are of different types.

The nuclear force also has a tensor component which depends on the interaction between the nucleon spins and the angular momentum of the nucleons, leading to deformation from a simple spherical shape.

Nuclear Binding

To disassemble a nucleus into unbound protons and neutrons requires work against the nuclear force. Conversely, energy is released when a nucleus is created from free nucleons or other nuclei: the nuclear binding energy. Because of mass–energy equivalence (i.e. Einstein's famous formula $E = mc^2$), releasing this energy causes the mass of the nucleus to be lower than the total mass of the individual nucleons, leading to the so-called "mass defect".

The nuclear force is nearly independent of whether the nucleons are neutrons or protons. This property is called *charge independence.* The force depends on whether the spins of the nucleons are parallel or antiparallel, as it has a non-central or *tensor* component. This part of the force does not conserve orbital angular momentum, which under the action of central forces is conserved.

The symmetry resulting in the strong force, proposed by Werner Heisenberg, is that protons and neutrons are identical in every respect, other than their charge. This is not completely true, because neutrons are a tiny bit heavier, but it is an approximate symmetry. Protons and neutrons are therefore viewed as the same particle, but with different isospin quantum numbers; conventionally, the proton is *isospin up,* while the neutron is *isospin down.* The strong force is invariant under SU(2) isospin transformations, just as other interactions between particles are invariant under SU(2) transformations of intrinsic spin. In other words, both isospin and intrinsic spin transformations are isomorphic to the SU(2) symmetry group. There are only strong attractions when the total isospin of the set of interacting particles is 0, which is confirmed by experiment.

Our understanding of the nuclear force is obtained by scattering experiments and the binding energy of light nuclei. The nuclear force occurs by the exchange of virtual light mesons, such as the virtual pions, as well as two types of virtual mesons with spin (vector mesons), the rho mesons and the omega mesons. The vector mesons account for the spin-dependence of the nuclear force in this "virtual meson" picture.

The nuclear force is distinct from what historically was known as the weak nuclear force. The weak interaction is one of the four fundamental interactions, and plays a role in such processes as beta decay. The weak force plays no role in the interaction of nucleons, though it is responsible for the decay of neutrons to protons and vice versa.

Nuclear Force as a Residual of the Strong Force

The nuclear force is a residual effect of the more fundamental strong force, or strong interaction. The strong interaction is the attractive force that binds the elementary particles called quarks together to form the nucleons (protons and neutrons) themselves. This more powerful force is mediated by particles called gluons. Gluons hold quarks together through color charge which is analogous to electric charge, but far stronger. Quarks, gluons, and their dynamics are mostly confined within nucleons, but residual influences extend slightly beyond nucleon boundaries to give rise to the nuclear force.

The nuclear forces arising between nucleons are analogous to the forces in chemistry between neutral atoms or molecules called London forces. Such forces between atoms are much weaker than the attractive electrical forces that hold the atoms themselves together (i.e., that bind electrons to the nucleus), and their range between atoms is shorter, because they arise from small separation of charges inside the neutral atom. Similarly, even though nucleons are made of quarks in combinations which cancel most gluon forces (they are "color neutral"), some combinations of quarks and gluons nevertheless leak away from nucleons, in the form of short-range nuclear force fields that extend from one nucleon to another nearby nucleon. These nuclear forces are very weak compared to direct gluon forces ("color forces" or strong forces) inside nucleons, and the nuclear forces extend only over a few nuclear diameters, falling exponentially with distance. Nevertheless, they are strong enough to bind neutrons and protons over short distances, and overcome the electrical repulsion between protons in the nucleus.

Sometimes, the nuclear force is called the residual strong force, in contrast to the strong interactions which arise from QCD. This phrasing arose during the 1970s when QCD was being established. Before that time, the *strong nuclear force* referred to the inter-nucleon potential. After the verification of the quark model, *strong interaction* has come to mean QCD.

Nucleon–nucleon Potentials

Two-nucleon systems such as the deuteron, the nucleus of a deuterium atom, as well as proton–proton or neutron–proton scattering are ideal for studying the *NN* force. Such systems can be described by attributing a *potential* (such as the Yukawa potential) to the nucleons and using the potentials in a Schrödinger equation. The form of the potential is derived phenomenologically (by measurement), although for the long-range interaction, meson-exchange theories help to construct the potential. The parameters of the potential are determined by fitting to experimental data such as the deuteron binding energy or *NN* elastic scattering cross sections (or, equivalently in this context, so-called *NN* phase shifts).

The most widely used *NN* potentials are the Paris potential, the Argonne AV18 potential , the CD-Bonn potential and the Nijmegen potentials.

A more recent approach is to develop effective field theories for a consistent description of nucleon–nucleon and three-nucleon forces. Quantum hadrodynamics is an effective field theory of the nuclear force, comparable to QCD for color interactions and QED for electromagnetic interactions. Additionally, chiral symmetry breaking can be analyzed in terms of an effective field theory (called chiral perturbation theory) which allows perturbative calculations of the interactions between nucleons with pions as exchange particles.

From Nucleons to Nuclei

The ultimate goal of nuclear physics would be to describe all nuclear interactions from the basic interactions between nucleons. This is called the *microscopic* or *ab initio* approach of nuclear physics. There are two major obstacles to overcome before this dream can become reality:

- Calculations in many-body systems are difficult and require advanced computation techniques.
- There is evidence that three-nucleon forces (and possibly higher multi-particle interactions) play a significant role. This means that three-nucleon potentials must be included into the model.

This is an active area of research with ongoing advances in computational techniques leading to better first-principles calculations of the nuclear shell structure. Two- and three-nucleon potentials have been implemented for nuclides up to $A=12$.

Nuclear Potentials

A successful way of describing nuclear interactions is to construct one potential for the whole nucleus instead of considering all its nucleon components. This is called the *macroscopic* approach. For example, scattering of neutrons from nuclei can be described by considering a plane wave in the potential of the nucleus, which comprises a real part and an imaginary part. This model is often called the optical model since it resembles the case of light scattered by an opaque glass sphere.

Nuclear potentials can be *local* or *global*: local potentials are limited to a narrow energy range and a narrow nuclear mass range, while global potentials, which have more parameters and are usually less accurate, are functions of the energy and the nuclear mass and can therefore be used in a wider range of applications.

Nucleons

Nucleon is the collective term for protons and neutrons. Nucleons are the particles found in the nucleus of atoms. Most notably nucleons are a result of the strong force holding the atoms together—which is stronger than the electric force pushing them apart. Through a beta decay, the weak force can turn one nucleon into another nucleon—either protons to neutrons or neutrons to protons. Nucleons are incredibly small, about 10^{-15} m, 10,000x smaller than an atom.

Electrons are not nucleons, therefore, don't feel the strong force. The strong force only acts on objects made of quarks.

Mesons

In particle physics, mesons are hadronic subatomic particles composed of one quark and one antiquark, bound together by strong interactions. Because mesons are composed of quark subparticles, they have physical size, notably a diameter of roughly one femtometer, which is about 1.2 times the size of a proton or neutron. All mesons are unstable, with the longest-lived lasting for only a few hundredths of a microsecond. Charged mesons decay (sometimes through mediating particles) to form electrons and neutrinos. Uncharged mesons may decay to photons. Both of these decays imply that color is no longer a property of the byproducts.

Outside the nucleus, mesons appear in nature only as short-lived products of very high-energy collisions between particles made of quarks, such as cosmic rays (high-energy protons and neutrons) and ordinary matter. Mesons are also frequently produced artificially in cyclotron in the collisions of protons, antiprotons, or other particles.

Higher energy (more massive) mesons were created momentarily in the Big Bang, but are not thought to play a role in nature today. However, such heavy mesons are regularly created in particle accelerator experiments, in order to understand the nature of the heavier types of quark that compose the heavier mesons.

Mesons are part of the hadron particle family, and are defined simply as particles composed of an even number of quarks. The other members of the hadron family are the baryons: subatomic particles composed of odd numbers of valence quarks (at least 3), and some experiments show evidence of exotic mesons, which do not have the conventional valence quark content of two quarks (one quark and one antiquark), but 4 or more.

Because quarks have a spin of $\frac{1}{2}$, the difference in quark number between mesons and baryons results in conventional two-quark mesons being bosons, whereas baryons are fermions.

Each type of meson has a corresponding antiparticle (antimeson) in which quarks are replaced by their corresponding antiquarks and vice versa. For example, a positive pion (π^+) is made of one up quark and one down antiquark; and its corresponding antiparticle, the negative pion (π^-), is made of one up antiquark and one down quark.

Because mesons are composed of quarks, they participate in both the weak and strong interactions. Mesons with net electric charge also participate in the electromagnetic interaction. Mesons are classified according to their quark content, total angular momentum, parity and various other properties, such as C-parity and G-parity. Although no meson is stable, those of lower mass are nonetheless more stable than the more massive, and hence are easier to observe and study in particle accelerators or in cosmic ray experiments. Mesons are also typically less massive than baryons, meaning that they are more easily produced in experiments, and thus exhibit certain higher-energy phenomena more readily than do baryons. For example, the charm quark was first seen in the J/Psi meson (J/ψ) in 1974, and the bottom quark in the upsilon meson (Υ) in 1977.

Spin, Orbital Angular Momentum and Total Angular Momentum

Spin (quantum number S) is a vector quantity that represents the "intrinsic" angular momentum of a particle. It comes in increments of $\frac{1}{2}\,\hbar$. The $\hbar$ is often dropped because it is the "fundamental" unit of spin, and it is implied that "spin 1" means "spin 1 $\hbar$". (In some systems of natural units, $\hbar$ is chosen to be 1, and therefore does not appear in equations).

Quarks are fermions—specifically in this case, particles having spin $\frac{1}{2}\left(S=\frac{1}{2}\right)$. Because spin projections vary in increments of 1 (that is 1 $\hbar$), a single quark has a spin vector of length $\frac{1}{2}$

and has two spin projections $\left(S_z = +{}^1\!/_2 and S_z = -{}^1\!/_2\right)$. Two quarks can have their spins aligned, in which case the two spin vectors add to make a vector of length S = 1 and three spin projections $\left(S_z = +{}^1\!/_2 and S_z = -{}^1\!/_2\right)$. called the spin-1 triplet. If two quarks have unaligned spins, the spin vectors add up to make a vector of length $S = 0$ and only one spin projection $\left(S_z = 0\right)$, called the spin-0 singlet. Because mesons are made of one quark and one antiquark, they can be found in triplet and singlet spin states.

There is another quantity of quantized angular momentum, called the orbital angular momentum (quantum number L), that comes in increments of 1 $\hbar$, which represent the angular momentum due to quarks orbiting around each other. The total angular momentum (quantum number J) of a particle is therefore the combination of intrinsic angular momentum (spin) and orbital angular momentum. It can take any value from $J = |L - S| to J = |L + S|$ in increments of 1.

Meson angular momentum quantum numbers for L = 0, 1, 2, 3				
S	L	J	P	J^P
0	0	0	−	0^-
	1	1	+	1^+
	2	2	−	2^-
	3	3	+	3^+
1	0	1	−	1^-
	1	2, 1, 0	+	$2^+, 1^+, 0^+$
	2	3, 2, 1	−	$3^-, 2^-, 1^-$
	3	4, 3, 2	+	$4^+, 3^+, 2^+$

Particle physicists are most interested in mesons with no orbital angular momentum (L = 0), therefore the two groups of mesons most studied are the S = 1; L = 0 and S = 0; L = 0, which corresponds to J = 1 and J = 0, although they are not the only ones. It is also possible to obtain J = 1 particles from S = 0 and L = 1. How to distinguish between the S = 1, L = 0 and S = 0, L = 1 mesons is an active area of research in meson spectroscopy.

Parity

If the universe were reflected in a mirror, most of the laws of physics would be identical—things would behave the same way regardless of what we call "left" and what we call "right". This concept of mirror reflection is called parity (P). Gravity, the electromagnetic force, and the strong interaction all behave in the same way regardless of whether or not the universe is reflected in a mirror, and thus are said to conserve parity (P-symmetry). However, the weak interaction does distinguish "left" from "right", a phenomenon called parity violation (P-violation).

Based on this, one might think that, if the wavefunction for each particle (more precisely, the quantum field for each particle type) were simultaneously mirror-reversed, then the new set of wavefunctions would perfectly satisfy the laws of physics (apart from the weak interaction). It turns out that this is not quite true: In order for the equations to be satisfied, the wavefunctions of certain types of particles have to be multiplied by −1, in addition to being mirror-reversed. Such particle types are said to have *negative* or *odd* parity ($P = -1$, or alternatively $P = -$whereas the other particles are said to have *positive* or *even* parity ($P=+1$, *or alternatively* $P=+$).

For mesons, the parity is related to the orbital angular momentum by the relation:

$$P = (-1)^{L+1}$$

where the L is a result of the parity of the corresponding spherical harmonic of the wavefunction. The "+ 1" comes from the fact that, according to the Dirac equation, a quark and an antiquark have opposite intrinsic parities. Therefore, the intrinsic parity of a meson is the product of the intrinsic parities of the quark (+1) and antiquark (−1). As these are different, their product is −1, and so it contributes the "+ 1" that appears in the exponent.

As a consequence, all mesons with no orbital angular momentum ($L = 0$) have odd parity ($P = -1$).

C-parity

C-parity is only defined for mesons that are their own antiparticle (i.e. neutral mesons). It represents whether or not the wavefunction of the meson remains the same under the interchange of their quark with their antiquark. If

$$|q\bar{q}\rangle = |\bar{q}q\rangle$$

then, the meson is "C even" (C = +1). On the other hand, if

$$|q\bar{q}\rangle = -|\bar{q}q\rangle$$

then the meson is "C odd" (C = −1).

C-parity rarely is studied on its own, but more commonly in combination with P-parity into CP-parity. CP-parity was thought to be conserved, but was later found to be violated in weak interactions.

G-parity

G parity is a generalization of the C-parity. Instead of simply comparing the wavefunction after exchanging quarks and antiquarks, it compares the wavefunction after exchanging the meson for the corresponding antimeson, regardless of quark content.

If

$$|q_1\overline{q_2}\rangle = |\overline{q_1}q_2\rangle|$$

then, the meson is "G *even*" (G = + 1) On the other hand, if

$$|q_1\overline{q_2}\rangle = -|\overline{q_1}q_2\rangle$$

then the meson is "*G odd*" ($G=-1$).

Isospin and Charge

The concept of isospin was first proposed by Werner Heisenberg in 1932 to explain the similarities between protons and neutrons under the strong interaction. Although they had different electric charges, their masses were so similar that physicists believed that they were actually the same particle. The different electric charges were explained as being the result of some unknown excitation similar to spin. This unknown excitation was later dubbed *isospin* by Eugene Wigner in 1937. When the first mesons were discovered, they too were seen through the eyes of isospin and so the three pions were believed to be the same particle, but in different isospin states.

This belief lasted until Murray Gell-Mann proposed the quark model in 1964 (containing originally only the u, d, and s quarks). The success of the isospin model is now understood to be the result of the similar masses of the u and d quarks. Because the u and d quarks have similar masses, particles made of the same number of them also have similar masses. The exact specific u and d quark composition determines the charge, because u quarks carry charge $+\frac{2}{3}$ whereas d quarks carry charge $-\frac{1}{3}$. For example, the three pions all have different charges $\pi^+(u\bar{d}), \pi^0$ (a quantum superposition of $u\bar{u}$ and $d\bar{d}$ states), $\pi^-(u\bar{d}), \pi^0$ but have similar masses (c. 140 MeV/c^2) as they are each made of a same number of total of up and down quarks and antiquarks. Under the isospin model, they were considered to be a single particle in different charged states.

The mathematics of isospin was modeled after that of spin. Isospin projections varied in increments of 1 just like those of spin, and to each projection was associated a "charged state". Because the "pion particle" had three "charged states", it was said to be of isospin $I = 1$. Its "charged states" $\pi^+, \pi^0, and\, \pi^-$, corresponded to the isospin projections $I_3 = +1, I_3 = 0, and\, I_3 = -1$ respectively. Another example is the "rho particle", also with three charged states. Its "charged states" ρ^+, ρ^0, *and* ρ^-, corresponded to the isospin projections $I_3 = +1$, $I_3 = 0$, and $I_3 = -1$respectively. It was later noted that the isospin projections were related to the up and down quark content of particles by the relation

$$I_3 = \frac{1}{2}[(n_u - n_{\bar{u}}) - (n_d - n_{\bar{d}})],$$

where the *n*'s are the number of up and down quarks and antiquarks.

In the "isospin picture", the three pions and three rhos were thought to be the different states of two particles. However, in the quark model, the rhos are excited states of pions. Isospin, although conveying an inaccurate picture of things, is still used to classify hadrons, leading to unnatural and often confusing nomenclature. Because mesons are hadrons, the isospin classification is also used, with $I_3 = +\frac{1}{2}$ for up quarks and down antiquarks, and $I_3 = -\frac{1}{2}$ for up antiquarks and down quarks.

Flavour Quantum Numbers

The strangeness quantum number *S* was noticed to go up and down along with particle mass. The higher the mass, the lower the strangeness (the more s quarks). Particles could be described with isospin projections (related to charge) and strangeness (mass)). As other quarks were discovered, new quantum numbers were made to have similar description of udc and udb nonets. Because only the u and d mass are similar, this description of particle mass and charge in terms of isospin and flavour quantum numbers only works well for the nonets made of one u, one d and one other quark and breaks down for the other nonets (for example ucb nonet). If the quarks all had the same mass, their behaviour would be called *symmetric*, because they would all behave in exactly the same way with respect to the strong interaction. However, as quarks do not have the same mass, they do not interact in the same way (exactly like an electron placed in an electric field will accelerate more than a proton placed in the same field because of its lighter mass), and the symmetry is said to be broken.

It was noted that charge (Q) was related to the isospin projection (I_3) the baryon number (B) and flavour quantum numbers (S,C,B',T) by the Gell-Mann–Nishijima formula:

$$Q = I_3 + \frac{1}{2}(B + S + C + B' + T),$$

where S, C, B', and T represent the strangeness, charm, bottomness and topness flavour quantum numbers respectively. They are related to the number of strange, charm, bottom, and top quarks and antiquark according to the relations:

$$S = -(n_s - n_{\bar{s}})$$
$$C = +(n_c - n_{\bar{c}})$$
$$B' = -(n_b - n_{\bar{b}})$$
$$T = +(n_t - n_{\bar{t}}),$$

meaning that the Gell-Mann–Nishijima formula is equivalent to the expression of charge in terms of quark content:

$$Q = \frac{2}{3}[(n_u - n_{\bar{u}}) + (n_c - n_{\bar{c}}) + (n_t - n_{\bar{t}})] - \frac{1}{3}[(n_d - n_{\bar{d}}) + (n_s - n_{\bar{s}}) + (n_b - n_{\bar{b}})].$$

Classification

Mesons are classified into groups according to their isospin (I), total angular momentum (J), parity (P), G-parity (G) or C-parity (C) when applicable, and quark (q) content. The rules for classification are defined by the Particle Data Group, and are rather convoluted. The rules are presented below, in table form for simplicity.

Types of Meson

Mesons are classified into types according to their spin configurations. Some specific configurations are given special names based on the mathematical properties of their spin configuration.

Types of mesons					
Type	***S***	***L***	***P***	***J***	J^P
Pseudoscalar meson	0	0	–	0	0^-
Pseudovector meson	0, 1	1	+	1	1^+

Vector meson	1	0, 2	–	1	1^-
Scalar meson	1	1	+	0	0^+
Tensor meson	1	1, 3	+	2	2^+

Nomenclature

Flavourless Mesons

Flavourless mesons are mesons made of pair of quark and antiquarks of the same flavour (all their flavour quantum numbers are zero: $S= 0, C= 0, B'= 0, T= 0$). The rules for flavourless mesons are:

Nomenclature of Flavourless Mesons

$q\bar{q}$ **content**	$J^{PC^{\dagger}} \rightarrow I$	0^{-+}, 2^{-+}, 4^{-+}, ...	1^{+-}, 3^{+-}, 5^{+-},...	1^{--}, 2^{--}, 3^{--}, ...	0^{++}, 1^{++}, 2^{++}, ...
ud $\frac{u\bar{u}-d\bar{d}}{\sqrt{2}}$ du	1	π^+ π^0 π^-	$b^+ b^0 b^-$	ρ^+ ρ^0 ρ^-	$a^+ a^0 a^-$
$u\bar{u}$ Mix of $d\bar{d}$ $s\bar{s}$	0	η η'	h h′	ω ϕ	f f′
$c\bar{c}$	0	η_c	h_c	$\psi^{\dagger\dagger}$	χ_c
$b\bar{b}$	0	η_c	h_b	Υ	χ_b
$t\bar{t}$	0	η_t	h_t	θ	χ_t

† ^ The C parity is only relevant to neutral mesons.

†† ^ For $J^{PC}=1^{--}$, the ψ is called the J/ψ

In addition:

- When the spectroscopic state of the meson is known, it is added in parentheses.
- When the spectroscopic state is unknown, mass (in MeV/c^2) is added in parentheses.
- When the meson is in its ground state, nothing is added in parentheses.

Flavoured Mesons

Flavoured mesons are mesons made of pair of quark and antiquarks of different flavours. The rules are simpler in this case: the main symbol depends on the heavier quark, the superscript depends on the charge, and the subscript (if any) depends on the lighter quark. In table form, they are:

Nomenclature of flavoured mesons						
Antiquark → quark ↓	Up	Down	Charm	Strange	Top	Bottom
Up	—	[21]	D^0	K^+	T^0	B^+
Down	[21]	—	D^- -	K^0	T^-	B^0 o
Charm	D^0	D^+	—	D^+_s s	T^0_c c	B^+_c c
Strange	K^-	K^0 o	D^-_s s	—	T^-_s s	B^0_s
Top	T^0	T^+	T^0_c c	T^+_s s	—	T^+_b b
Bottom	B^-	B^0	B^-_c	B^0_s s	T^-_b	—

In addition:

- If J^P is in the "normal series" (*i.e.*, $J^P = 0^+, 1^-, 2^+, 3^-, \ldots$) a superscript $*$ is added.
- If the meson is not pseudoscalar ($J^P = 0^-$) or vector ($J^P = 1^-$) J is added as a subscript.
- When the spectroscopic state of the meson is known, it is added in parentheses.
- When the spectroscopic state is unknown, mass (in MeV/c^2) is added in parentheses.
- When the meson is in its ground state, nothing is added in parentheses.

Exotic Mesons

There is experimental evidence for particles that are hadrons (i.e., are composed of quarks) and are color-neutral with zero baryon number, and thus by conventional definition are mesons. Yet, these particles do not consist of a single quark/antiquark pair. A tentative category for these particles is exotic mesons.

There are at least five exotic meson resonances that have been experimentally confirmed to exist by two or more independent experiments. The most statistically significant of these is the Z(4430), discovered by the Belle experiment in 2007 and confirmed by LHCb in 2014. It is a candidate for being a tetraquark: a particle composed of two quarks and two antiquarks.

Pseudoscalar Mesons

Particle name	Particle symbol	Antiparticle symbol	Quark content	Rest mass (MeV/c²)	I^G	J^{PC}	S	C	B'	Mean lifetime (s)	Commonly decays to (>5% of decays)
Pion	π^+	π^-	$u\bar{d}$	139.57018±0.00035	1^-	0^-	0	0	0	$(2.6033\pm0.0005)\times10^{-8}$	$\mu^+ + \nu_\mu$
Pion	π^0	Self	$\frac{u\bar{u}-d\bar{d}}{\sqrt{2}}$[a]	0134.9766±0.0006	1^-	0^{-+}	0	0	0	$(8.4\pm0.6)\times10^{-17}$	$\gamma+\gamma$
Eta meson	η	Self	$\frac{u\bar{u}+d\bar{d}-2s\bar{s}}{\sqrt{6}}$[a]	547.853±0.024	0^+	0^{-+}	0	0	0	$(5.0\pm0.3)\times10^{-19}$[b]	$\gamma+\gamma$ or $\pi^0+\pi^0+\pi^0$ or $\pi^+ + \pi^0 + \pi^-$
Eta prime meson	η' (958)	Self	$\frac{u\bar{u}+d\bar{d}+s\bar{s}}{\sqrt{3}}$[a]	957.66±0.24	0^+	0^{-+}	0	0	0	$(3.2\pm0.2)\times10^{-21}$[b]	$\pi^+ + \pi^- + \eta$ or $(\rho^0+\gamma)/(\pi^+ + \pi^- + \gamma)$ or $\pi^0+\pi^0+\eta$
Charmed eta meson	η'_c(1S)	Self	$c\bar{c}$	2980.3±1.2	0^+	0^{-+}	0	0	0	$(2.5\pm0.3)\times10^{-23}$[b]	
Bottom eta meson	η_b (1S)	Self	$b\bar{b}$	9300±40	0^+	0^{-+}	0	0	0	Unknown	
Kaon	K^+	K^-	$u\bar{s}$	493.677±0.016	½	0^-	1	0	0	$(1.2380\pm0.0021)\times10^{-8}$	$\mu^+ + \nu_\mu$ or $\pi^+ + \pi^0$ or $\pi^0 + e^+ + \nu_e$ or $\pi^+ + \pi^0$
Kaon	K^0	$\overline{K}^0$	$d\bar{s}$	497.614±0.024	½	0^-	1	0	0	[c]	[c]
K-Short	$K^0{}_S$	Self	$\frac{d\bar{s}-s\bar{d}}{\sqrt{2}}$]	497.614±0.024[d]	½	0^-	(*)	0	0	$(8.953\pm0.005)\times10^{-11}$	$\pi^+ + \pi^-$ or $\pi^0+\pi^0$
K-Long	$K^0{}_L$	Self	$\frac{d\bar{s}+s\bar{d}}{\sqrt{2}}$ [e]]	497.614±0.024[d]	½	0^-	(*)	0	0	$(5.116\pm0.020)\times10^{-8}$	$\pi^\pm + e^\mp + \nu_e$ or $\pi^\pm + \mu^\mp + \nu_\mu$ or $\pi+\pi^0+\pi^0$ or $\pi^+ + \pi^0 + \pi^-$
D meson	D^+	D^-	$c\bar{d}$	1869.62±0.20	½	0^-	0	+1	0	$(1.040\pm0.007)\times10^{-12}$	
D meson	D^0	$\overline{D}^0$	$c\bar{u}$	1864.84±0.17	½	0^-	0	+1	0	$(4.101\pm0.015)\times10^{-13}$	
Strange D meson		$D^-{}_s$	$c\bar{s}$	1968.49±0.34	0	0^-	+1	+1	0	$(5.00\pm0.07)\times10^{-13}$	

B meson	B^+	B^-	$u\bar{b}$	5279.15±0.31	½	0^-	0	0	+1	$(1.638±0.011)×10^{-12}$	
B meson	B^0 0	$\overline{B}^0$	$d\bar{b}$	5279.53±33	½	0^-	0	0	+1	$(1.530±0.009)×10^{-12}$	
Strange B meson	$B^0{}_s$	$B^0{}_s$	$s\bar{b}$	5366.3±0.6	0	0^-	−1	0	+1	1.470+0.026 −0.027×10^{-12}	
Charmed B meson	$B^+{}_c$	$B^-{}_c$	$c\bar{b}$	6276±4	0	0^-	0	+1	+1	$(4.6±0.7)×10^{-13}$	

- Makeup inexact due to non-zero quark masses.
- PDG reports the resonance width (Γ). Here the conversion $\tau = {}^{\hbar}/_{\Gamma}$ is given instead.
- Strong eigenstate. No definite lifetime.
- Tha mass of the $K^0{}_L$ and $K^0{}_S$ are given as that of the K^0 however it is known that a difference between the masses of the $K^0{}_L$ and $K^0{}_S$ on the order of 2.2×10^{-11} MeV / c exists.
- Weak eigenstate. Makeup is missing small CP–violating term.

Vector Mesons

Particle name	Particle symbol	Antiparticle symbol	Quark content	Rest mass(MeV/c^2)	I^G	J^{PC}	S	C	B'	Mean lifetime (s)	Commonly decays to (>5% of decays)
Charged rho meson	ρ^+ (770)	ρ^- (770)	$u\bar{d}$	775.4±0.4	1^+	1^-	0	0	0	~$4.5×10^{-24}$[f][g]	$\pi^\pm + \pi^0$
Neutral rho meson	ρ^0 (770)	Self	$\frac{u\bar{u}-d\bar{d}}{\sqrt{2}}$	775.49±0.34	1^+	1^{--}	0	0	0	~$4.5×10^{-24}$[f][g]	$\pi^+ + \pi^-$
Omega meson	ω (782)	Self	$\frac{u\bar{u}+d\bar{d}}{\sqrt{2}}$	782.65±0.12	0^-	1^{--}	0	0	0	$(7.75±0.07)×10^{-23}$[f]	$\pi^+ + \pi^0 + \pi^-$ or $\pi^0 + \gamma$
Phi meson	ϕ (1020)	Self	$s\bar{s}$	1019.445±0.020	0^-	1^{--}	0	0	0	$(1.55±0.01)×10^{-22}$[f]	$K^+ + K^-$ or $K^0{}_S + K^0{}_L$ or $(\rho + \pi) / (\pi^+ + \pi^0 + \pi^-)$
J/Psi	J/ψ	Self	$c\bar{c}$	3096.916±0.011	0^-	1^{--}	0	0	0	$(7.1±0.2)×10^{-21}$[f]	
Upsilon meson	Υ (1S)	Self	$b\bar{b}$	9460.30±0.26	0^-	1^{--}	0	0	0	$(1.22±0.03)×10^{-20}$[f]	
Kaon	K^{*+}	K^{*-}	$u\bar{s}$	891.66±0.026	½	1^-	1	0	0	~$7.35×10^{-20}$[f][g]	
Kaon	K^{*0}	K^{*0}	$d\bar{s}$	896.00±0.025	½	1^-	1	0	0	$(7.346±0.002)×10^{-20}$[f]	
D meson	D^{*+} (2010)	D^{*-} (2010)	$c\bar{d}$	2010.27±0.17	½	1^-	0	+1	0	$(6.9±1.9)×10^{-21}$[f]	$D^0 + \pi^+$ or $D^+ + \pi^0$
D meson	D^{*0} (2007)	D^{*0} (2007)	$c\bar{u}$	2006.97±0.19	½	1^-	0	+1	0	$>3.1×10^{-22}$[f]	$D^0 + \pi^0$ or $D^0 + \gamma$
strange D meson	$D^{*+}{}_s$	$D^{*-}{}_s$	$c\bar{s}$	2112.3±0.5	0	1^-	+1	+1	0	$>3.4×10^{-22}$[f]	$D^{*+} + \gamma$ or $D^{*+} + \pi^0$
B meson	B^{*+}	B^{*-}	$u\bar{b}$	5325.1±0.5	½	1^-	0	0	+1	Unknown	$B^+ + \gamma$

B meson	B^{*0}	B^{*0}	$d\bar{b}$	5325.1±0.5	½	1^-	0	0	+1	Unknown	$B^0 + \gamma$
Strange B meson	B^{*0}_s	B^{*0}_s	$s\bar{b}$	5412.8±1.3	0	1^-	−1	0	+1	Unknown	$B^0_s + \gamma$
Charmed B meson†	B^{*+}_c	B^{*-}_c	$c\bar{b}$	Unknown	0	1^-	0	+1	+1	Unknown	Unknown

- PDG reports the resonance width (Γ). Here the conversion $\tau = \hbar/_{\Gamma}$ is given instead.
- The exact value depends on the method used.

Notes on Neutral Kaons

There are two complications with neutral kaons:

- Due to neutral kaon mixing, the K^0_S *and* K^0_L are not eigenstates of strangeness. However, they *are* eigenstates of the weak force, which determines how they decay, so these are the particles with definite lifetime.
- The linear combinations given in the table for the K^0_S *and* K^0_L are not exactly correct, since there is a small correction due to CP violation.

Note that these issues also exist in principle for other neutral flavored mesons; however, the weak eigenstates are considered separate particles only for kaons because of their dramatically different lifetimes.

Isospin

In physics, and specifically, particle physics, isospin (isotopic spin, isobaric spin) is a symmetry of the strong interaction as it applies to the interactions of the neutron and proton. Isospin symmetry is a subset of the flavour symmetry seen more broadly in the interactions of baryons and mesons. Isospin symmetry remains an important concept in particle physics, and a close examination of this symmetry historically led directly to the discovery and understanding of quarks and of the development of Yang-Mills theory.

Symmetry

Isospin was introduced by Werner Heisenberg to explain several related symmetries:

- The mass of the neutron and the proton are almost identical: they are nearly degenerate, and are thus often called nucleons. Although the proton has a positive charge, and the neutron is neutral, they are almost identical in all other respects.
- The strength of the strong interaction between any pair of nucleons is the same, independent of whether they are interacting as protons or as neutrons.
- The mass of the pions which mediate the strong interaction between the nucleons are the same. In particular, the mass of the positively-charged pion is identical to that of the negatively-charged pion, and both have nearly the same mass as the neutral pion.

In quantum mechanics, when a Hamiltonian has a symmetry, that symmetry manifests itself through a set of states that have (almost) the same energy; that is, the states are degenerate. In particle physics, mass is the same thing as energy (since $E=mc^2$), and so the near mass-degeneracy of the neutron and proton points at a symmetry of the Hamiltonian describing the strong interactions. The neutron does have a slightly higher mass: the mass degeneracy is not exact. The proton is charged, the neutron is not. However, here, as the case would be in general for quantum mechanics, the appearance of a symmetry can be imperfect, as it is perturbed by other forces, which give rise to slight differences between states.

SU(2)

Heisenberg's contribution was to note that the mathematical formulation of this symmetry was in certain respects similar to the mathematical formulation of spin, from whence the name "isospin" derives. To be precise, the isospin symmetry is given by the invariance of the Hamiltonian of the strong interactions under the action of the Lie group $SU(2)$. The neutron and the proton are assigned to the doublet (the spin-1/2 or fundamental representation) of $SU(2)$. The pions are assigned to the triplet (the spin-1 or adjoint representation) of $SU(2)$.

Just as is the case for regular spin, isospin is described by two numbers, I, the total isospin, and I_3 the component of the spin vector in a given direction. The proton and neutron both have $I=1/2$ as they belong to the doublet. The proton has $I_3=+1/2$ or 'isospin-up' and the neutron has $I_3=-1/2$ or 'isospin-down'. The pions, belonging to the triplet, have $I=1$, *and* π^+, π^0 and π^- have, respectively, $I_3=+1,\ 0,\ -1$.

Yang-Mills

Isospin symmetry was central to the original formulation of Yang-Mills theory. The pions were proposed to be the SU(2) gauge bosons of this theory. While it is now understood that isospin symmetry is not a true gauge symmetry, this initial confusion was historically important for the development of the overall ideas of gauge invariance.

Relationship to Flavour

The discovery and subsequent close analysis of additional particles, both mesons and baryons, made it clear that the concept of isospin symmetry could be broadened to an even larger symmetry group, now called flavour symmetry. Once the kaons and their property of strangeness became better understood, it started to become clear that these, too, seemed to be a part of an enlarged, more general symmetry that contained isospin as a subset. The larger symmetry was named the Eight-fold Way by Murray Gell-Mann, and was promptly recognized to correspond to the adjoint representation of SU(3). This immediately led to Gell-Mann's proposal of the existence of quarks. The quarks would belong to the fundamental representation of the flavour SU(3) symmetry, and it is from the fundamental rep and its conjugate (the quarks and the anti-quarks) that the higher representation (the mesons and baryons) could be assembled. In short, the theory of Lie groups and Lie algebras modelled the physical reality of particles in the most exceptional and unexpected way.

The discovery of the J/ψ meson and charm led to the expansion of flavour symmetry to SU(4), and the discovery of the upsilon meson (and the corresponding top and bottom quarks) led to the current SU(6) flavour symmetry. Isospin symmetry is just one little corner of this broader symmetry. There are strong theoretical reasons, confirmed by experiment, that lead one to believe that things stop there, and that there are no further quarks to be found.

Isospin Symmetry of Quarks

In the framework of the Standard Model, the isospin symmetry of the proton and neutron are reinterpreted as the isospin symmetry of the up and down quarks. Technically, the nucleon doublet is seen to be the product of a single quark (thus, a doublet) and a pair of quarks in a singlet state. That is, the proton wave function, in terms of quark flavour eigenstates, is described by

$$|p\rangle = |u\rangle \frac{1}{\sqrt{2}}(|ud\rangle + |du\rangle) + \text{perms.}$$

and the neutron by

$$|n\rangle = |d\rangle \frac{1}{\sqrt{2}}(|ud\rangle + |du\rangle) + \text{perms.}$$

where perms stands for permutations. Here, $|u\rangle$ is the up quark flavour eigenstate, and $|d\rangle$ is the down quark flavour eigenstate. Although the above is the technically correct way of denoting a proton and neutron in terms of quark flavour eigenstates, this is almost always glossed over, and these are more simply referred to as uud and udd.

Similarly, the isopsin symmetry of the pions are given by:

$$|\pi^+\rangle = |u\bar{d}\rangle$$
$$|\pi^0\rangle = \frac{1}{\sqrt{2}}\left(|u\bar{u}\rangle - |d\bar{d}\rangle\right)$$
$$|\pi^-\rangle = |d\bar{u}\rangle$$

The overline denotes, as usual, the complex conjugate representation of SU(2), or, equivalently, the antiquark.

Weak Isospin

The quarks also feel the weak interaction; however, the mass eigenstates of the strong interaction are not exactly the same as the eigenstates of the weak interaction. Thus, while there are still a pair of quarks u and d that take part in the weak interaction, they are not quite the same as the strong u and d quarks. The difference is given by a rotation, whose magnitude is called the Cabibbo angle or more generally, the CKM matrix.

Hypercharge

In particle physics is a quantum number relating the strong interactions of the SU(3) model.

Isospin is defined in the SU(2) model while the SU(3) model defines hypercharge.

SU(3) weight diagrams are 2-dimensional with the coordinates referring to two quantum numbers, I_z, which is the *z*-component of isospin and *Y*, which is the hypercharge (the sum of strangeness *S*, charm *C*, bottomness *B′*, topness *T*, and baryon number *B*). Mathematically, hypercharge is:

$$Y = S + C + B' + T + B$$

and conservation of hypercharge implies a conservation of flavour. Strong interactions conserve hypercharge, but weak interactions do not.

Relation with Electric Charge and Isospin

The Gell-Mann–Nishijima formula relates isospin and electric charge,

$$Q = I_3 + \tfrac{1}{2}Y,$$

where I_3 is the third component of isospin and *Q* is the particle's charge.

Isospin creates multiplets of particles whose average charge is related to the hypercharge by:

$$Y = 2\bar{Q}.$$

since the hypercharge is the same for all members of a multiplet, and the average of the I_3 values is 0.

SU(3) Model in Relation to Hypercharge

The SU(2) model has multiplets characterized by a quantum number J, which is the total angular momentum. Each multiplet consists of $2J + 1$substates with equally-spaced values of J_z forming a symmetric arrangement seen in atomic spectra and isospin. This formalizes the observation that certain strong baryon decays were not observed, leading to the prediction of the mass, strangeness and charge of the Ω^-baryon.

The SU(3) has *supermultiplets* containing SU(2) multiplets. SU(3) now needs two numbers to specify all its sub-states which are denoted by λ_1 *and* λ_2.

$(\lambda_1 + 1)$ specifies the number of points in the topmost side of the hexagon while $(\lambda_2 + 1)$ specifies the number of points on the bottom side.

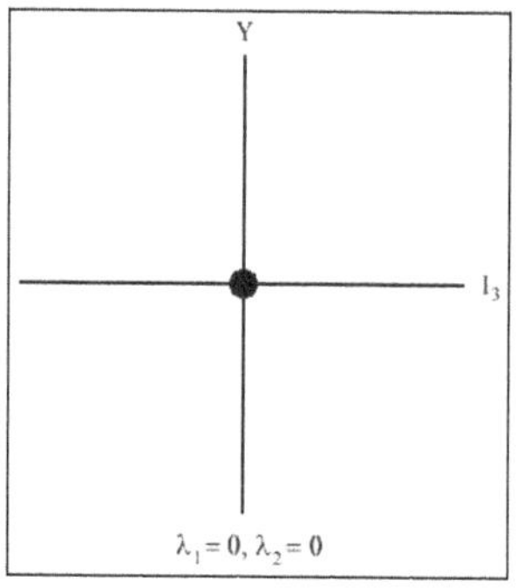

SU(3) singlet weight diagram, where Y is hypercharge and I_3 is the third component of isospin.

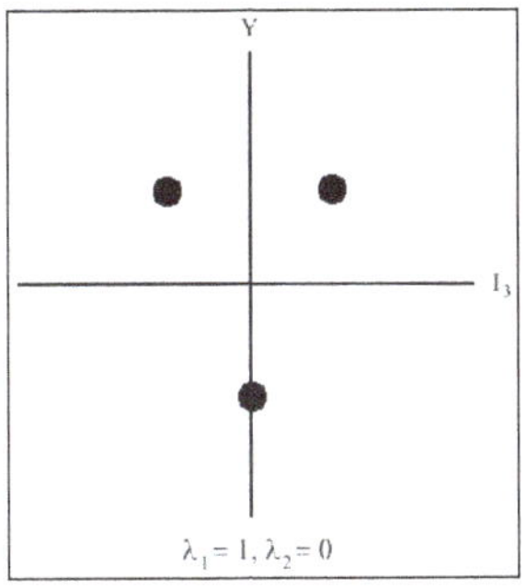

SU(3) triplet weight diagram

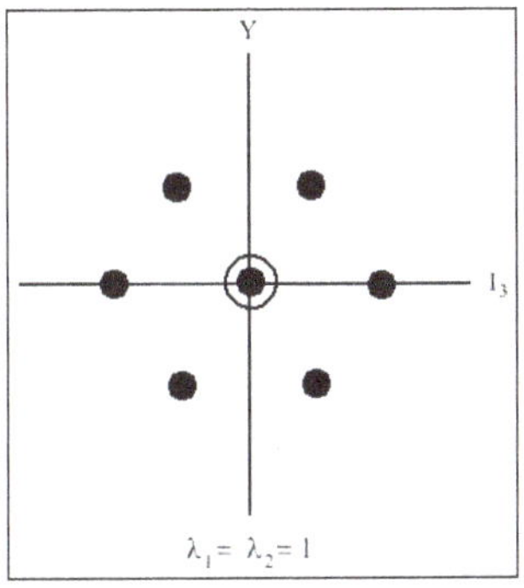

SU(3) septet, octet, and nonet weight diagram Note similarity with both charts on the right. The number used to describe the weight diagram depends on whether the particle(s) occupying the center of the diagram have one, two, or three distinct names.

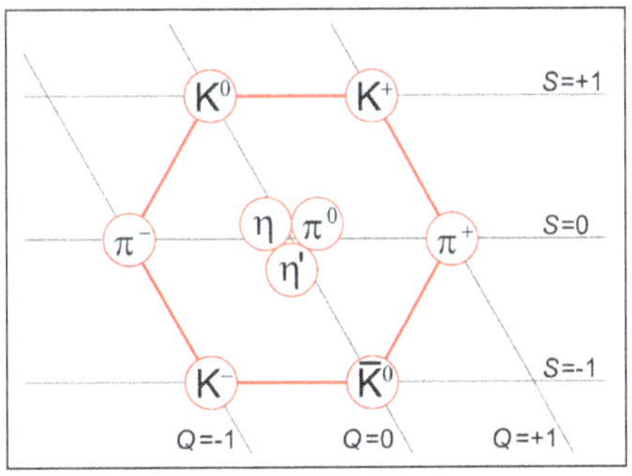

Mesons of spin 0 form a nonet. K: kaon, π: pion, η: eta meson.

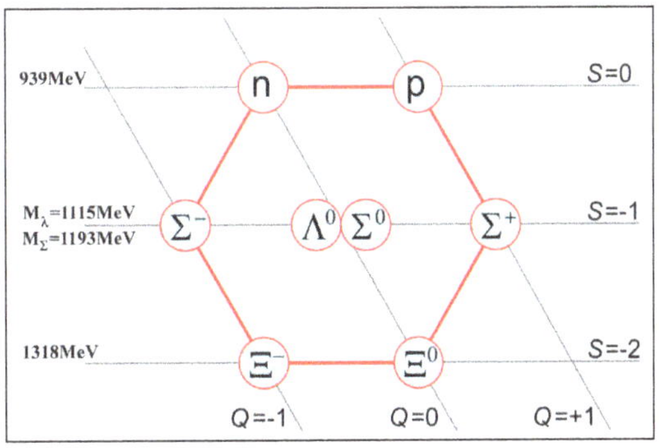

The octet of light spin-$\frac{1}{2}$ baryons described in SU(3). n: neutron, p: proton, Λ: Lambda baryon, Σ: Sigma baryon, Ξ: Xi baryon.

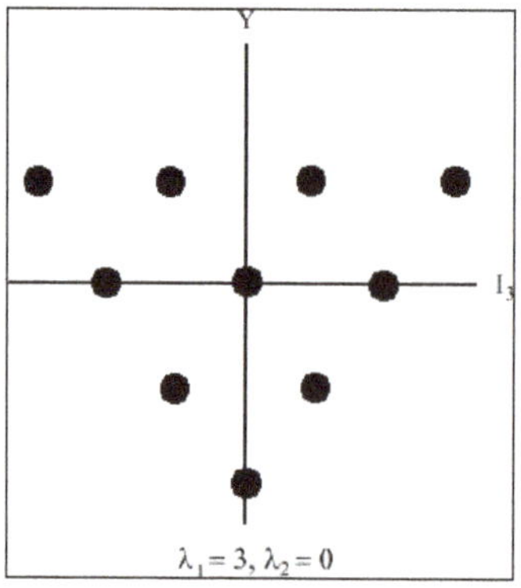

SU(3) decuplet weight diagram Note similarity with chart on the right.

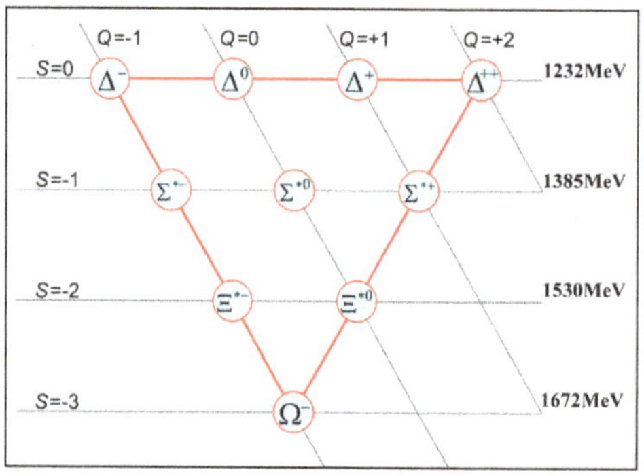

A combination of three up, down or strange quarks with a total spin of $\frac{3}{2}$ form the so-called baryon decuplet. The lower six are hyperons. *S*: strangeness, *Q*: electric charge.

Examples:

- The nucleon group (protons with Q=+1 and neutrons with Q=0 have an average charge of $+\frac{1}{2}$ so they both have hypercharge $Y = 1$ (baryon number B=+1,S=C=B'=T=0) From the Gell-Mann–Nishijima formula we know that proton has isospin $I_3 = +\frac{1}{2}$, while neutron has $I_3 = -\frac{1}{2}$.
- This also works for quarks: for the *up* quark, with a charge of $+\frac{2}{3}$, and an I_3 of we deduce a hypercharge of $\frac{1}{3}$, due to its baryon number (since three quarks make a baryon, a quark has a baryon number $\frac{1}{3}$).
- For a *strange* quark, with charge $-\frac{1}{3}$, a baryon number of $\frac{1}{3}$ and strangeness of −1 we get a hypercharge $Y = -2/3$, so we deduce an I_3=0. That means that a *strange* quark makes an isospin singlet of its own (same happens with *charm*, *bottom* and *top* quarks), while *up* and *down* constitute an isospin doublet.

Practical Obsolescence

Hypercharge was a concept developed in the 1960s, to organize groups of particles in the *"particle zoo"* and to develop *ad hoc* conservation laws based on their observed transformations. With the advent of the quark model, it is now obvious that hypercharge Y is the following combination of the numbers of up (n_u), down (n_d), strange (n_s), charm (n_c), top (n_t) and bottom (n_b):

$$Y = \tfrac{1}{3}\left(n_u + n_d - 2n_s + 4n_c - 2n_b + 4n_t\right).$$

In modern descriptions of hadron interaction, it has become more obvious to draw Feynman diagrams that trace through individual quarks composing the interacting baryons and mesons, rather than counting hypercharge quantum numbers. Weak hypercharge, however, remains of practical use in various theories of the electroweak interaction.

Baryons

The term baryon usually refers to a subatomic particle composed of three quarks. A more technical (and broader) definition is that it is a subatomic particle with a baryon number of 1. Baryons are a subset of hadrons, (which are particles made of quarks), and they participate in the strong interaction. They are also a subset of fermions. Well-known examples of baryons are protons and neutrons, which make up atomic nuclei, but many unstable baryons have been found as well.

Some "exotic" baryons, known as pentaquarks, are thought to be composed of four quarks and one antiquark, but their existence is not generally accepted. Each baryon has a corresponding antiparticle, called an anti-baryon, in which quarks are replaced by their corresponding antiquarks.

Basic Properties

Each baryon has an odd half-integer spin (such as ${}^{1}/_{2}\, or\, {}^{3}/_{2}$), where "spin" refers to the angular momentum quantum number. Baryons are therefore classified as fermions. They experience the strong nuclear force and are described by Fermi-Dirac statistics, which apply to all particles obeying the Pauli exclusion principle. This stands in contrast to bosons, which do not obey the exclusion principle.

Baryons, along with mesons, are hadrons, meaning they are particles composed of quarks. Each quark has a baryon number of $B = -{}^{1}/_{3}$, and each antiquark has a baryon number of:

$$B = -{}^{1}/_{3}.$$

The term baryon number is defined as:

$$B = \frac{N_q - N_{\bar{q}}}{3}$$

where,

N_q is the number of quarks,

$N_{\bar{q}}$ is the number of antiquarks.

The term "baryon" is usually used for triquarks, that is, baryons made of three quarks. Thus, each baryon has a baryon number of 1 ($B={}^1/_3+{}^1/_3+{}^1/_3=1$).

Some have suggested the existence of other, "exotic" baryons, such as pentaquarks—baryons made of four quarks and one antiquark ($B={}^1/_3+{}^1/_3+{}^1/_3+{}^1/_3-{}^1/_3=1$))—but their existence is not generally accepted. Theoretically, heptaquarks (5 quarks, 2 antiquarks), nonaquarks (6 quarks, 3 antiquarks), and so forth could also exist.

Besides being associated with a spin number and a baryon number, each baryon has a quantum number known as strangeness. This quantity is equal to -1 times the number of strange quarks present in the baryon.

Classification

Baryons are classified into groups according to their isospin values and quark content. There are six groups of triquarks:

- Nucleon (N),
- Delta (Δ),
- Lambda (Λ),
- Sigma (Σ),
- Xi (Ξ),
- Omega (Ω).

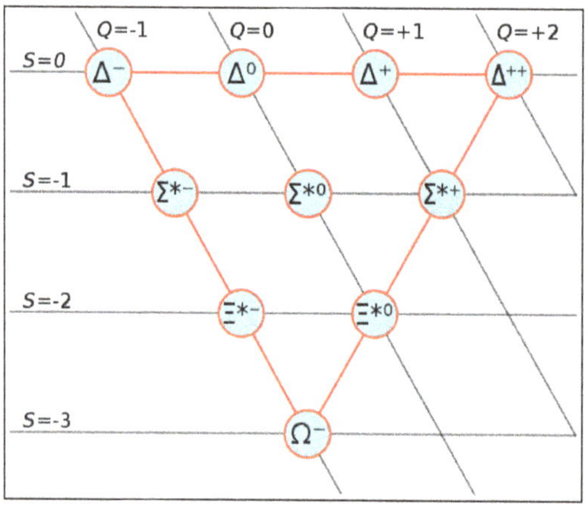

Combinations of three u, d or s-quarks with a total spin of 3/2 form the so-called baryon decuplet.

The rules for classification are defined by the Particle Data Group. The rules cover all the particles that can be made from three of each of the six quarks (up, down, strange, charm, bottom, top), although baryons made of top quarks are not expected to exist because of the top quark's short lifetime. (The rules do not cover pentaquarks.) According to these rules, the u, d, and s quarks are considered light, and the c, b, and t quarks are considered heavy.

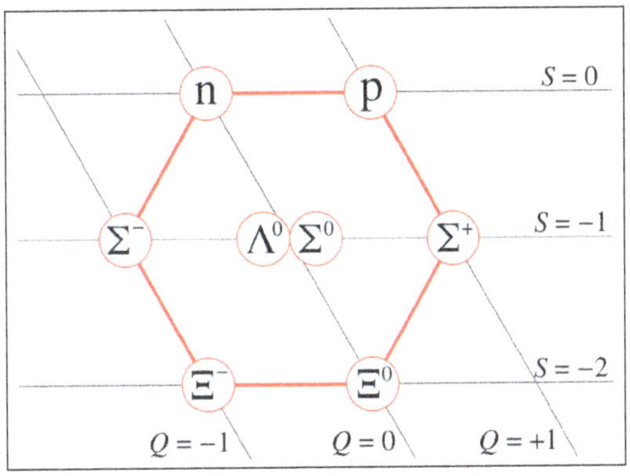

The octet of light spin-1/2 baryons.

Based on the rules, the following classification system has been set up:

- Baryons with three u and d quarks are grouped as N Ξ (*isospin* $^1/_2$) or Δ (*isospin* 3/2).
- Baryons with two u and d quarks are grouped as Λ (isospin 0) or Σ (isospin 1). If the third quark is heavy, its identity is given by a subscript.
- Baryons with one u or d quark are placed in the group Ξ (*isospin* $^1/_2$) One or two subscripts are used if one or both of the remaining quarks are heavy.
- Baryons with no u or d quarks are placed in the group (Ω) (isospin 0), and subscripts indicate any heavy quark content.
- Some baryons decay strongly, in which case their masses are shown as part of their names. For example, Sigmas (Σ) and Omegas (Ω) do not decay strongly, but Deltas (Δ(1232)) and charmed *Xis* (Ξ_c^+(2645) do.

Given that quarks carry charge, knowledge of the charge of a particle indirectly gives the quark content. For example, the rules say that the Σ_b contains a bottom and some combination of two up and down quarks. $A\Sigma_b^0$ must be one up quark $(Q=-^2/_3)$ one down quark $(Q=-^1/_3)$, and one bottom quark $(Q=-^1/_3)$) to have the correct charge $(Q=0)$.

The number of baryons within one group (excluding resonances) is given by the number of isospin projections possible (2 × isospin + 1). For example, there are four Δ's, corresponding to the four isospin projections of the isospin value $I={}^3/_2:\Delta^{++}\ (I_z={}^3/_2),\Delta^+(I_z={}^1/_2),\Delta^0(I_z=-{}^1/_2)$, and $\Delta^-(I_z=-{}^3/_2)$. Another example would be the three Σ_b's corresponding to the three isospin projections of the isospin value $I=1:\Sigma_b^+(I_z=1),\Sigma_b^0(I_z=0),\ and \Sigma_b^-(I_z=-1).$).

Charmed Baryons

Baryons that are composed of at least one charm quark are known as charmed baryons.

Baryonic Matter

Baryonic matter is matter composed mostly of baryons (by mass). It includes atoms of all types, and thus includes nearly all types of matter that we may encounter or experience in everyday life, including the matter that constitutes human bodies. Non-baryonic matter, as implied by the name, is any sort of matter that is not primarily composed of baryons. It may include such ordinary matter as neutrinos or free electrons, but it may also include exotic species of non-baryonic dark matter, such as supersymmetric particles, axions, or black holes.

The distinction between baryonic and non-baryonic matter is important in cosmology, because Big Bang nucleosynthesis models set tight constraints on the amount of baryonic matter present in the early universe.

The very existence of baryons is also a significant issue in cosmology because current theory assumes that the Big Bang produced a state with equal amounts of baryons and anti-baryons. The process by which baryons came to outnumber their antiparticles is called baryogenesis. (This is distinct from a process by which leptons account for the predominance of matter over antimatter, known as leptogenesis).

Baryogenesis

Experiments are consistent with the number of quarks in the universe being a constant and, more specifically, the number of baryons being a constant; in technical language, the total baryon number appears to be conserved. Within the prevailing Standard Model of particle physics, the number of baryons may change in multiples of three due to the action of sphalerons, although this is rare and has not been observed experimentally. Some grand unified theories of particle physics also predict that a single proton can decay, changing the baryon number by one; however, this has not yet been observed experimentally. The excess of baryons over antibaryons in the present universe is thought to be due to non-conservation of baryon number in the very early universe, though this is not well understood.

Gluons

A gluon is an elementary particle that acts as the exchange particle (or gauge boson) for the strong force between quarks. It is analogous to the exchange of photons in the electromagnetic force between two charged particles. In layman's terms, they "glue" quarks together, forming hadrons such as protons and neutrons.

In technical terms, gluons are vector gauge bosons that mediate strong interactions of quarks in quantum chromodynamics (QCD). Gluons themselves carry the color charge of the strong interaction. This is unlike the photon, which mediates the electromagnetic interaction but lacks an electric charge. Gluons therefore participate in the strong interaction in addition to mediating it, making QCD significantly harder to analyze than QED (quantum electrodynamics).

Counting Gluons

Unlike the single photon of QED or the three W and Z bosons of the weak interaction, there are eight independent types of gluon in QCD.

This may be difficult to understand intuitively. Quarks carry three types of color charge; antiquarks carry three types of anticolor. Gluons may be thought of as carrying both color and anticolor. This gives nine *possible* combinations of color and anticolor in gluons. The following is a list of those combinations (and their schematic names):

- Red-antired ($r\bar{r}$), red-antigreen ($r\bar{g}$), red-antiblue ($r\bar{b}$),
- Green-antired ($g\bar{r}$), green-antigreen ($g\bar{g}$), green-antiblue ($g\bar{b}$),
- Blue-antired, (($b\bar{r}$)), blue-antigreen ($b\bar{g}b\bar{g}$), blue-antiblue ($b\bar{b}$).

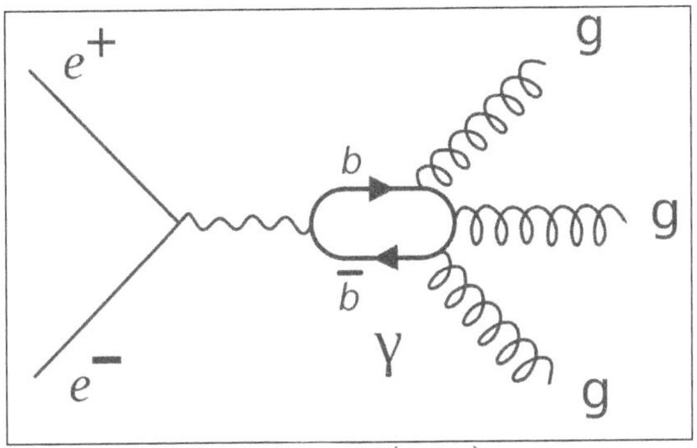

Diagram: $e^+e^- \to \Upsilon(9.46) \to 3g$

These are not the *actual* color states of observed gluons, but rather *effective* states. To correctly understand how they are combined, it is necessary to consider the mathematics of color charge in more detail.

Color Singlet States

It is often said that the stable strongly interacting particles (such as the proton and the neutron, i.e. hadrons) observed in nature are "colorless", but more precisely they are in a "color singlet" state, which is mathematically analogous to a *spin* singlet state. Such states allow interaction with other color singlets, but not with other color states; because long-range gluon interactions do not exist, this illustrates that gluons in the singlet state do not exist either.

The color singlet state is:

$$(r\bar{r} + b\bar{b} + g\bar{g})/\sqrt{3}.$$

In words, if one could measure the color of the state, there would be equal probabilities of it being red-antired, blue-antiblue, or green-antigreen.

Eight Gluon Colors

There are eight remaining independent color states, which correspond to the "eight types" or "eight colors" of gluons. Because states can be mixed together as discussed above, there are many ways of presenting these states, which are known as the "color octet". One commonly used list is:

$$(r\bar{b} + b\bar{r})/\sqrt{2} \qquad -i(r\bar{b} - b\bar{r})/\sqrt{2}$$
$$(r\bar{g} + g\bar{r})/\sqrt{2} \qquad -i(r\bar{g} - g\bar{r})/\sqrt{2}$$
$$(b\bar{g} + g\bar{b})/\sqrt{2} \qquad -i(b\bar{g} - g\bar{b})/\sqrt{2}$$
$$(r\bar{r} - b\bar{b})/\sqrt{2} \qquad (r\bar{r} + b\bar{b} - 2g\bar{g})/\sqrt{6}.$$

These are equivalent to the Gell-Mann matrices. The critical feature of these particular eight states is that they are linearly independent, and also independent of the singlet state, hence $3^2 - 1$ *or* 2^3. There is no way to add any combination of these states to produce any other, and it is also impossible to add them to make rr, gg, or bb the forbidden singlet state. There are many other possible choices, but all are mathematically equivalent, at least equally complicated, and give the same physical results.

Group Theory Details

Technically, QCD is a gauge theory with SU(3) gauge symmetry. Quarks are introduced as spinors in N_f flavors, each in the fundamental representation (triplet, denoted **3**) of the color gauge group, SU(3). The gluons are vectors in the adjoint representation (octets, denoted **8**) of color SU(3). For a general gauge group, the number of force-carriers (like photons or gluons) is always equal to the dimension of the adjoint representation. For the simple case of SU(*N*), the dimension of this representation is $N^2 - 1$.

In terms of group theory, the assertion that there are no color singlet gluons is simply the statement that quantum chromodynamics has an SU(3) rather than a U(3) symmetry. There is no known *a priori* reason for one group to be preferred over the other, but as discussed above, the experimental evidence supports SU(3). The U(1) group for electromagnetic field combines with a slightly more complicated group known as SU(2) – S stands for "special" – which means the corresponding matrices have determinant +1 in addition to being unitary.

Confinement

Since gluons themselves carry color charge, they participate in strong interactions. These gluon-gluon interactions constrain color fields to string-like objects called "flux tubes", which exert constant force when stretched. Due to this force, quarks are confined within composite particles called hadrons. This effectively limits the range of the strong interaction to 1×10^{-15} meters, roughly the size of an atomic nucleus. Beyond a certain distance, the energy of the flux tube binding two quarks increases linearly. At a large enough distance, it becomes energetically more favorable to pull a quark-antiquark pair out of the vacuum rather than increase the length of the flux tube.

Gluons also share this property of being confined within hadrons. One consequence is that gluons are not directly involved in the nuclear forces between hadrons. The force mediators for these are other hadrons called mesons.

Although in the normal phase of QCD single gluons may not travel freely, it is predicted that there exist hadrons that are formed entirely of gluons — called glueballs. There are also conjectures about other exotic hadrons in which real gluons (as opposed to virtual ones found in ordinary hadrons) would be primary constituents. Beyond the normal phase of QCD (at extreme temperatures and pressures), quark–gluon plasma forms. In such a plasma there are no hadrons; quarks and gluons become free particles.

Experimental Observations

Quarks and gluons (colored) manifest themselves by fragmenting into more quarks and gluons, which in turn hadronize into normal (colorless) particles, correlated in jets. As shown in 1978 summer conferences, the PLUTO detector at the electron-positron collider DORIS (DESY) produced the first evidence that the hadronic decays of the very narrow resonance Υ(9.46) could be interpreted as three-jet event topologies produced by three gluons. Later, published analyses by the same experiment confirmed this interpretation and also the spin 1 nature of the gluon.

In summer 1979, at higher energies at the electron-positron collider PETRA (DESY), again three-jet topologies were observed, now interpreted as qq gluon bremsstrahlung, now clearly visible, by TASSO, MARK-J and (later in 1980 also by JADE). The spin 1 of the gluon was confirmed in 1980 by TASSO and). In 1991 a subsequent experiment at the LEP storage ring at CERN again confirmed this result.

The gluons play an important role in the elementary strong interactions between quarks and gluons, described by QCD and studied particularly at the electron-proton collider HERA at DESY. The number and momentum distribution of the gluons in the proton (gluon density) have been measured by two experiments, H1 and ZEUS, in the years 1996-2007. The gluon contribution to the proton spin has been studied by the HERMES experiment at HERA. The gluon density in the proton (when behaving hadronically) also has been measured.

Color confinement is verified by the failure of free quark searches (searches of fractional charges). Quarks are normally produced in pairs (quark + antiquark) to compensate the quantum color and flavor numbers; however at Fermilab single production of top quarks has been shown (technically this still involves a pair production, but quark and antiquark are of different flavor). No glueball has been demonstrated.

Deconfinement was claimed in 2000 at CERN SPS in heavy-ion collisions, and it implies a new state of matter: quark–gluon plasma, less interacting than in the nucleus, almost as in a liquid. It was found at the Relativistic Heavy Ion Collider (RHIC) at Brookhaven in the years 2004–2010 by four contemporaneous experiments. A quark–gluon plasma state has been confirmed at the CERN Large Hadron Collider (LHC) by the three experiments ALICE, ATLAS and CMS in 2010.

References

- Strong-force, science: britannica.com, Retrieved 18 June, 2019
- Povh, B.; Rith, K.; Scholz, C.; Zetsche, F. (2002). Particles and Nuclei: An Introduction to the Physical Concepts. Berlin: Springer-Verlag. p. 73. ISBN 978-3-540-43823-6
- Nucleon, encyclopedia: energyeducation.ca, Retrieved 25 July, 2019
- C. Amsler (Particle Data Group); et al. (2008). "Review of Particle Physics" (PDF). Physics Letters B. 667 (1): 1–1340. Bibcode:2008PhLB..667....1A. doi:10.1016/j.physletb.2008.07.018
- Baryon: newworldencyclopedia.org, Retrieved 19 August, 2019
- P. Söding (2010). "On the discovery of the gluon". European Physical Journal H. 35 (1): 3–28. Bibcode:2010EPJH...35....3S. doi:10.1140/epjh/e2010-00002-5

Weak Interactions 6

- **Free Neutron Decay**
- **Electroweak Theory**
- **Weak Isospin**
- **Weak Hypercharge**

The mechanism of interaction between subatomic particles responsible for the radioactive decay of atoms is known as weak interaction. It is primarily understood in terms of electroweak theory and consists of properties like weak hypercharge and weak isospin. This chapter has been carefully written to provide an easy understanding of weak interactions.

The weak interaction is responsible for radioactive decays. It is characterised by long lifetimes, and small cross sections. All fermions feel the weak interaction. When present, though, strong and electromagnetic interactions dominate.

Neutrinos feel only the weak interaction, which is what makes them so difficult to study. They are the only particles to experience just one of the fundamental forces.

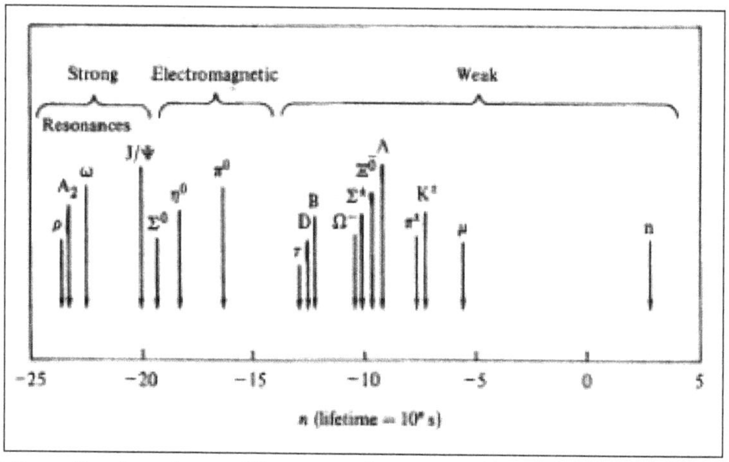

Lifetime of various decays. The strong decays are the fastest, followed by the electromagnetic decays and then the weak decays.

The 4-point Interaction

He first attempt to construct a theory of the weak interaction was made by Fermi in 1932. In analogy to the electromagnetic interaction, he imagined a 4-point interaction that happened at a single point in space-time. His idea of β-decay is shown in the figure.

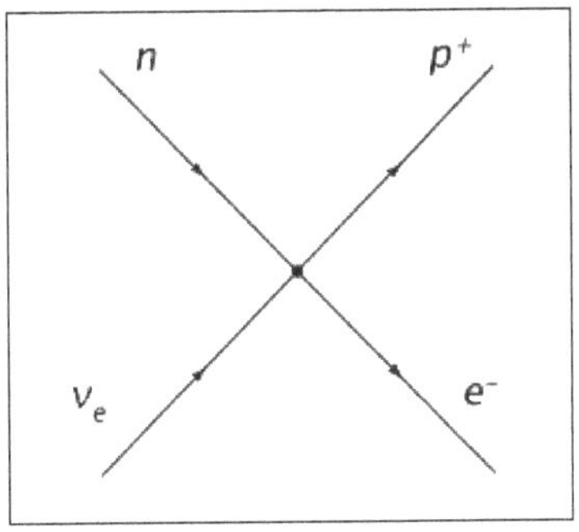

Fermi's 4-point interaction.

In analogy to the electromagnetic interaction, Fermi proposed the following matrix element,

$$M = \frac{G_F}{\sqrt{2}}\left[\overline{uP}\gamma^{\mu}uN\right]\left[u_e\gamma^{\nu}u_{\nu}\right].$$

We take note of some points here:

- Charge changing: The hadronic current has $\triangle Q = +1$, whereas the lepton current has $\triangle Q = +1$. There is net charge transferred from the hadronic to the lepton current and so we all this a charged current interaction.
- Universality: There is a coupling factor, GF , called the Fermi Constant, equal to 1.166 × 10−5GeV−2. Fermi postulated, and is has later been shown to stand up to experiment, that the weak coupling factor is the same for all weak vertices, regardless of the flavour of lepton taking part. This is called universality and is an extremely important concept.
- There is no propagator.
- The currents have a vector character, purely in analogy to the electromagnetic interaction where it was known that the currents were vector in nature.

The cross section for the interaction $\nu_e + n \rightarrow p + e^-$ as generated from Fermi's 4-point interaction, was calculate shortly after by Bethe. He found that

$$\sigma\left(n + \nu_e \rightarrow e^- + p\right) \sim E_{\nu}\left(M_e V\right) \times 10^{-43} cm^2$$

This is extremely small. You would need about 50 light-years of water to stop one 1 MeV neutrino. This cross-section also has a problem. It rises linearly with energy for ever. This is clearly incorrect and shows that the Fermi model breaks down at high energies. We need a bit of modification to the theory. We need to add a propagator.

Weak Propagator

We now know that the weak interaction is mediated by two massive gauge bosons : the charged $W^{\pm}$ and the neutral Z^{0}. The propagation term for the massive boson is $\frac{1}{M^2_{W,Z} - q^2}$. If we assume that the Fermi theory is the low energy limit of the Weak Interaction, then we can estimate the intrinsic coupling at high energy. In the Fermi limit, the coupling factor appears to $\frac{G_F}{\sqrt{2}}$. At low energies, with $M^2_{W,Z} >> q^2$, the propagator term reduces to just $\frac{1}{M^2_W}$ and we can make the identification,

$$\frac{G_F}{\sqrt{2}} = \frac{g_w^{\ 2}}{8M^2_w}$$

Parity and The Parity Operator

The parity operation is defined as spatial inversion around the origin :

$$t' \equiv t \; x' \equiv -x \; y' \equiv -y \; z' \equiv -z$$

Consider a Dirac spinor $\psi(t, x, t, z)$ A parity transformation would transform this spinor to,

$$\hat{P}\psi(t, x, y, z) = \psi(t, -x, -y, -z) = \psi'(t', x', y', 'z')$$

We can prove that the relevant operator is actually γ^0. That is,

$$\psi'(t', x', y', z') = \psi(t, -x, -y, -z) = \pm\gamma^0\psi(t, x, y, z)$$

Consider a Dirac spinor, $\psi(t, x, y, z)$ that obeys the Dirac equation,

$$i\gamma^0\frac{\partial\psi}{\partial t} + i\gamma^1\frac{\partial\psi}{\partial x} + i\gamma^2\frac{\partial\psi}{\partial y} + i\gamma^3\frac{\partial\psi}{\partial z} - m\psi = 0$$

Under the parity transformation : $\psi'(x', y', z', t') = \hat{P}\psi(x,y,z,t) = \gamma^0\psi(x, y, z, t)$ Since $(\gamma^0)^2 = 1$ this implies that,

$$\psi(x, y, z, t) = \gamma^0\psi'(x', y', z', t')$$

Substituting this into the Dirac equation we have,

$$i\gamma^0\gamma^0\frac{\partial\psi'}{\partial t} + i\gamma^1\gamma^0\frac{\partial\psi'}{\partial x} + i\gamma^2\gamma o\frac{\partial\psi'}{\partial y} + i\gamma^3\gamma^0\frac{\partial\psi'}{\partial z} - m\gamma^0\psi' = 0$$

We use the chain rule to express the derivative in terms of the primed coordinate system e.g.,

$$\frac{\partial\psi'}{\partial x} = \frac{\partial x'}{\partial x}\frac{\partial\psi'}{\partial x'} = -\frac{\partial\psi'}{\partial x'}$$

since $x' = -x$ under parity. In the Dirac equation,

$$i\frac{\partial\psi'}{\partial t'} - i\gamma^0\gamma^1\frac{\partial\psi'}{\partial x'} - i\gamma^0\gamma^2\frac{\partial\psi'}{\partial y'} - i\gamma^0\gamma^3\frac{\partial\psi'}{\partial z'} - m\gamma^0\psi' = 0$$

and since γ^0 anticommutes with γ^i for $i = 1, 2, 3$,

$$i\frac{\partial\psi'}{\partial t'} - i\gamma^0\gamma^1\frac{\partial\psi'}{\partial x'} - i\gamma^0\gamma^2\frac{\partial\psi'}{\partial y'} - i\gamma^0\gamma^3\frac{\partial\psi'}{\partial z'} - m\gamma^0\psi' = 0$$

Multiplying on the left by γ^0, and recalling that $(\gamma^0)^2 = 1$, we then get,

$$i\gamma^0\frac{\partial\psi'}{\partial t'} + i\gamma^1\frac{\partial\psi'}{\partial x'} + i\gamma^2\frac{\partial\psi'}{\partial y'} + i\gamma^3\frac{\partial\psi'}{\partial z'} - m\psi' = 0$$

which is the Dirac equation in the primed coordinates. Hence, under parity transformations the Dirac equation is unchanged (as it should be) provided that the bispinors transform as,

$$\psi \rightarrow \hat{P}\psi = \pm\gamma^0\psi$$

If we apply the parity operator twice then we must return the original wavefunction : $\hat{P}^2 = \gamma^{02} = 1$.

The eigenvalues of the parity operator are, therefore, ±1.. Hadrons are eigenstates of $\hat{P}$. The parity of a fermion is opposite that of the anti-fermion, whereas the parity of a boson is the same as its antiboson. We arbitrarily take particles to have positive or "even" intrinsic parity, and the anti-particle (if a fermion) is said to have negative or "odd" parity. The parity of a combined system is the product of the parity of its constituent parts.

Parity Violation

In 1956, T.D. Lee and C.N. Yang were trying to solve a very puzzling problem called the $\tau - \theta$ problem. Two strange mesons, called the τ and the θ, appeared to be identical in every respect: mass, spin, charge etc. The problem was that the τ was observed to decay into three pions $\pi^+\pi^+\pi^-$ *or* $\pi^+\pi^0\pi^0$. The other one, the θ , decays into two pions $\pi^+\pi^0$. Both are spin zero particles of strangeness one. The analysis of the final state showed that the τ decays into a parity odd state, while the θ into a parity even state. This seems impossible if the two particles were the same. Lee and Yang, after studying this, pointed out in 1956 that maybe these two particles could be the same particle. Of course this would be possible only if the parity is not preserved in these decays. They examined carefully the available evidence for parity conservation, and concluded that there was a lot of evidence for parity conservation in the strong and the electromagnetic interactions, while there was none in the weak interaction. They further proposed various ways the parity (non) conservation could be tested experimentally in the weak interaction.

Almost immediately C.S. Wu devised and carried a beautiful experiment to test the possibility of parity violation in beta decay. She set up a system of Co60 atoms which all decayed via β emission to N i60. She aligned them in a magnetic field, so that all their spin vectors lined up and then let them decay, measuring the direction of the outgoing electron. If parity were conserved, she would expect to see electrons emitted isotropically. Why? Have a look at figure, the spin vector of the Cobalt atom, labelled as J in the diagram, points to the left in both this world and the parity transformed mirror world. It is an example of an axial vector which doesn't change direction under a co-ordinate inversion (another example is angular momentum : $L = r \times p$. Under a co-ordinate inversion $r \to -r$ and $p \to -p$, which leaves the angular momentum unchanged). Suppose an electron were to be omitted in the direction of the spin vector in this world.

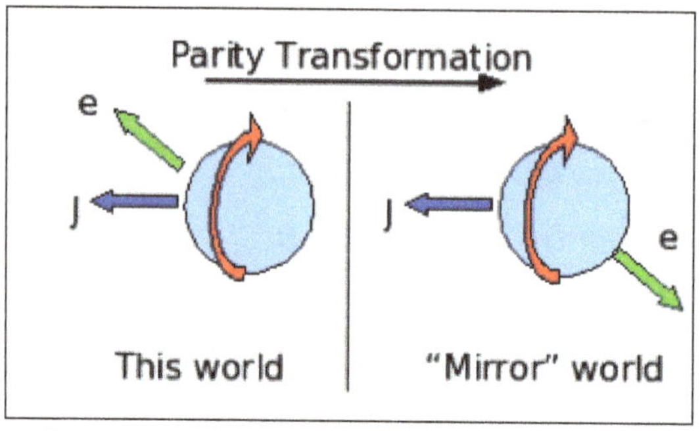

A schematic of Wu's parity conservation experiment.

In the mirror world the electron will be going in the other direction, opposite the direction of spin. Parity conservation implies that the probability of one interaction happening in this world is the same as the probability of it's mirror image occurring, and so we should see the same numbers of events wherethe electron were emitted anti-parallel to the spin, as the number of events in which the electron were emitted parallel to the spin vector.

What Wu saw was that electrons were emitted preferentially in the direction of the spin vector - a clear violation of parity conservation. It wasn't small either - almost all of the electrons were emitted in only one direction. It seemed as if the violation was maximal.

Parity, which had long been believed to be a true and fundamental symmetry of nature, fell in 1957, traumatising many respectable physicists.

CP Violation

Many desperate physicists tried to save the situation by appealing to CP invariance. We know that parity (P) is violated in the weak interaction, which can be seen from the decay

$$\pi^+ \rightarrow \mu^+ + \nu_\mu$$

in which the neutrino is always emitted with left-handed helicity. The weak interaction is not invariant under charge conjugation (C) either. For the charge conjugate of the previous decay is

$$\pi^- \rightarrow \mu^- + \bar{\nu}_\mu$$

In which the anti-neutrino still has left-handed helicity. The anti-neutrino in the real world always comes out right-handed. However if we combine the two operations we are back in business : CP changes a left-handed neutrino into a right-handed anti-neutrino, which is what is observed in nature. Many people breathed a sigh of relief, deciding that what we should have meant by the "mirror"-image of a right-handed electron was a left-handed positron.

Unhappily for them, CP is also violated. This was first shown by Cronin and Fitch in 1964. It's small, about 0.3% of weak interactions violate CP, but it's there. It means that there is a true violation of mirror symmetry in nature which can't be argued away be redefinitions, and that there is a difference in the laws of nature in our world and in the mirror world. This is lucky for us as it is probably the reason why we now live in a matter-dominated universe.

Table: All possible bilinear covariant combinations of γmatrices.

Name	Symbol	Current	Number of components	Effect under Parit
Scalar	S	$\overline{\psi}\psi$	1	+
Vector	V	$\overline{\psi}\gamma^{\mu}\psi$	4	(+,-,-,-)
Tensor	T	$\overline{\psi}\sigma^{\mu\nu}\psi$	6	
Axial Vector	A	$\overline{\psi}\gamma^{\mu}\gamma^{5}\psi$	4	(+,+,+,+)
Pseudo-Scalar	P	$\overline{\psi}\gamma^{5}\psi$	1	–

Building it into the Theory - the V-A Interaction

Alright. So parity is violated - let's not worry about how (in fact, noone really knows yet). How do we go about building this into our model so we can at least describe it? To do this we go back to our currents. The most general matrix element we can write is:

$$M \propto [\overline{u_{\psi,f}}\,\hat{O}\,u_{\psi,i}]\frac{1}{M^2 - q^2}[\overline{u_{\phi,f}}\,\hat{O}\,u_{\phi,i}]$$

where $\hat{O}$ is a combination of γ matrices.

It turns out that there are only 5 independent bilinear covariant expressions that you can form out of the γ matrices. They are labelled for how they behave under the Parity operation.

In this table $\sigma^{\mu\nu} = \frac{i}{2}\left(\gamma^{\mu}\gamma^{\nu} - \gamma^{\nu}\gamma^{\mu}\right)$.

Now, let's see how each of these currents behaves under a parity transformation. Ignoring the tensor current (which has two indices, rather than one and which therefore will not represent a theory which, at low energies, is a point-contact interaction) and noting that the parity transformation is:

$$\psi' = \gamma^{o}\psi$$

$$\overline{\psi'} = (\psi')^{\dagger\gamma^{o}} = (\gamma^{o}\psi)^{\dagger\gamma o} = \psi^{\dagger}\gamma^{o\dagger}\gamma^{o} = \psi^{\dagger}$$

where we have used the property that $\gamma^{o\dagger} = \gamma^{o}$ and $(\gamma^{o})^{2} = 1$.

- Scalar, S : $\bar{\psi}\psi \rightarrow \bar{\psi}\psi' = \psi'^{\dagger}\gamma^{0}\psi$

  $= \bar{\psi}\psi$

- Vector, V : $\bar{\psi}\gamma^{\mu}\psi \rightarrow \bar{\psi}\gamma^{\mu}\psi' = \bar{\psi}\gamma^{0}\gamma^{\mu}\gamma^{0}\psi$

  $= \bar{\psi}\gamma^{0}\psi\,(\mu = 0)$

  $-\bar{\psi}\gamma^{\mu}\psi\,(\mu > 0)$

- Axial Vector, A : $\bar{\psi}\gamma^{\mu}\gamma^{5}\psi \rightarrow \bar{\psi}'\gamma^{\mu}\gamma^{5}\psi' = \bar{\psi}\gamma^{0}\gamma^{\mu}\gamma^{5}\gamma^{0}\psi$

  $\bar{\psi}\gamma^{\mu}\gamma^{5}\psi$

- Pseudo Scalar, P : $\bar{\psi}\gamma^{5}\psi \rightarrow \bar{\psi}\gamma^{5}\psi' = \psi\gamma^{0}\gamma^{5}\gamma^{0}\psi$

  $= -\bar{\psi}\gamma^{5}\psi$

Unravelling which of these currents was responsible for the weak interaction took quite a lot of experimental and theoretical time. We are looking for a combination for which the charged weak interaction only couples to left-handed chiral particles. The left-handed chiral projection operator is $P_L = \frac{1}{2}(1-\gamma^5)$. Hence the current we want looks something like,

$$\bar{\psi}\,\hat{o}\,\frac{1}{2}(1-\gamma^5)\varphi$$

To cut a very long story short, experiment showed that the operator O^ was just the vector operator, γ^{μ}, so the whole interaction was,

$$\bar{\psi}\gamma^{\mu}\frac{1}{2}(1-\gamma^5)\varphi$$

If we expand this we get,

$$\frac{1}{2}\left(\bar{\psi}\gamma^{\mu}\varphi - \bar{\psi}\gamma^{\mu}\gamma^{5}\varphi\right)$$

and comparing to the table this makes the vector (V) and axial vector (A) currents responsible for the parity violating nature of the weak interaction. This is the famous **V-A** interaction. Parity violation comes from the fact that the behaviour of the vector and axial vector currents under a parity transformation are different. As you can see from the table, the vector current flips sign under parity whereas the axial vector doesn't. The interference between these two terms creates the parity violation. One can see this schematically by remembering that what we observe is usually the square of the amplitude. Suppose the amplitude is pure V-A. Then,

$$|M|^2 \sim (V-A)(V-A)$$

$$= VV - 2AV + AA$$

If we apply a parity transformation then the sign of the V term flips, but the sign of the A term doesn't.

$$\hat{P}\{|M|^2\} \sim \hat{P}\{(V-A)(V-A)\}$$
$$= \hat{P}\{VV - 2AV + AA\}$$
$$= (-V)(-V) + AA - 2A(-V)$$
$$= VV + AA + 2AV$$

Comparing the $|M|^2$ and $\hat{P}\{|M|^2\}$ we see a difference from $-2AV$ *to* $+2AV$. Without having the cross term, AV, made up of currents with opposite parity behaviours, one would end up with $|M|^2 = \hat{P}\{|M|^2\}$ and therefore there would be no parity violation.

The V-A interation actually violates parity maximally as both currents have the same strength. Parity isn't just violated in a small percentage of interactions, it's violated in all of them. One can test this by allowing the currents to have different weights,

$$\frac{1}{2}\overline{\psi}\gamma^{\mu}\left(cV - cA\gamma^5\right)\varphi$$

Experimentally it is found that $c_V = 1$ *and* $c_A = 1$.

The weak charged current can therefore be written as,

$$j^{CC}_{weak} = \frac{gw}{\sqrt{2}}\overline{u}\gamma^{\mu}\frac{1}{2}\left(1-\gamma^5\right)u$$

The V-A Interaction and Neutrinos

The inclusion of the left-handed chiral projection operator in the current implies that the charged weak interaction only couples left-handed chiral particles, or right-handed chiral antiparticles.

$$\overline{\psi}\gamma^{\mu}\frac{1}{2}\left(1-\gamma^5\right)\varphi = \left(\overline{\psi_L} + \overline{\psi_R}\right)\gamma^{\mu}\varphi L$$
$$= \overline{\psi_L}\gamma^{\mu}\varphi L$$

What does this mean for neutrinos? Well, we know that neutrinos are observed to all have lefthanded helicity, and anti-neutrinos all have right-handed helicity. Since neutrinos (even if they do have mass) are ultra-relativistic, this implies that all neutrinos have left-handed chirality, and antineutrinos have right-handed chirality. The neutrinos can only be made in weak interactions and so are all made as left-handed chiral particles. They have no choice.

This is an important but subtle point - neutrinos do not necessarily have intrinsic left-handed helicity. They have left-handed chirality because they can only be made by the weak interaction, and the weak interaction only makes left-handed chiral particles or right-handed chiral antiparticles. To a good approximation, since neutrinos are almost massless, helicity and chirality are the same thing, so the neutrino is always generated with left-handed helicity. This does not preclude the possibility of the existance of a neutrino with right-handed helicity. It can be shown, however, that the probability of generating a neutrino with right-handed helicity is proportional to $\left(\frac{m\nu}{E\nu}\right)^2$ and is therefore almost impossible (mv is the absolute neutrino mass. We know this is less than about $2\ eV$. For a neutrino with energy of, say, $10\ MeV$ the probability of emitting a wrong sign neutrino is around 4×10^{-14}).

This argument doesn't preclude the possibility of the existence of a right-handed chiral neutrino either. Unfortunately, if it does exist, it doesn't couple to any of our fundamental forces (with the possible exception of gravity, and even then it is extremely weak) and hence may as well not exist. Hence, the existence of a right-chiral neutrino state is almost a question of philosophy - but not quite.a right-chiral neutrino state could still have indirect but visible effects in some neutrino oscillation experiments.

Electrons, on the hand, are massive and can come in both left- and right-handed chiral states. However, only the left-handed electrons couple to the charged weak interaction, i.e. to the $W\pm$ boson. It is possible for the Z^0 to couple to right-handed chiral particles as well. As neutrinos are only created by the charged weak current, this makes no difference to the properties of the neutrino.

Free Neutron Decay

Outside the nucleus, free neutrons are unstable and have a mean lifetime of 881.5±1.5 s (about 14 minutes, 42 seconds). Therefore, the half-life for this process (which differs from the mean lifetime by a factor of ln(2) ≈ 0.693) is 611±1 s (about 10 minutes, 11 seconds). The beta decay of the neutron, described above, can be denoted as follows:

$$n^0 \rightarrow p^+ + e^- + \nu_e$$

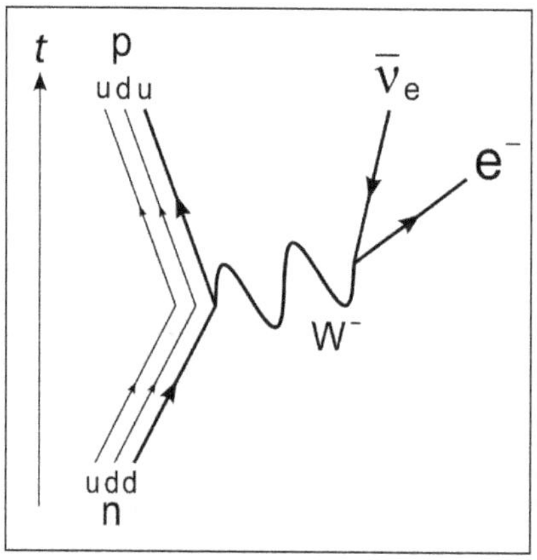

Feynman diagram for beta decay of the neutron.

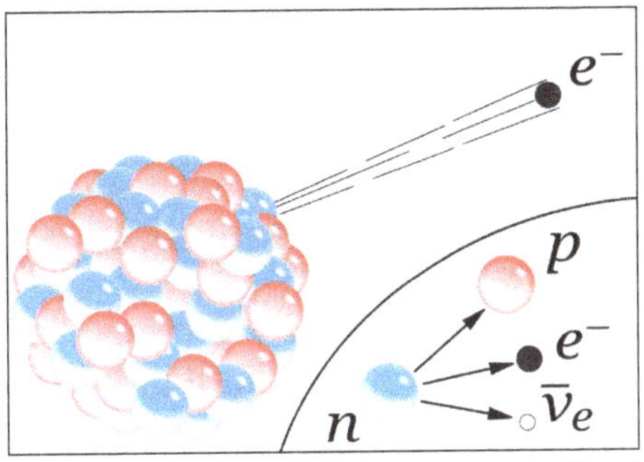

A schematic of the nucleus of an atom indicating.

β^-radiation, the emission of a fast electron from the nucleus (the accompanying antineutrino is omitted). In the Rutherford model for the nucleus, red spheres were protons with positive charge and blue spheres were protons tightly bound to an electron with no net charge.

The inset shows beta decay of a free neutron as it is understood today; an electron and antineutrino are created in this process.

This decay, like any flavor-changing process, occurs through operation of the weak force. It involves the emission of aW^- boson from one of the down quarks within the neutron, thereby converting the down quark into an up quark and the neutron into a proton; the W^- then decays into the electron and the antineutrino. The following equations denote the same process as the first equation above, but also include the short-lived $W^-$and describe the process on both the nucleon and the quark level:

$$n^0 \rightarrow p^+ + W^- \rightarrow p^+ + e^- + \nu_e$$

$$udd \rightarrow uud + W^- \rightarrow uud + e^- + \nu_e$$

For the free neutron, the decay energy for this process (based on the rest masses of the neutron, proton and electron) is 0.782343 MeV. That is the difference between the rest mass of the neutron and the sum of the rest masses of the products. That difference has to be carried away as kinetic energy. The maximal energy of the beta decay electron (in the process wherein the neutrino receives a vanishingly small amount of kinetic energy) has been measured at 0.782 ± .013 MeV. The latter number is not well-enough measured to determine the comparatively tiny rest mass of the neutrino (which must in theory be subtracted from the maximal electron kinetic energy); furthermore, neutrino mass is constrained by many other methods.

A small fraction (about one in 1000) of free neutrons decay with the same products, but add an extra particle in the form of an emitted gamma ray:

$$n^0 \rightarrow p^+ + e^- + \nu_e + \gamma$$

This gamma ray may be thought of as a sort of "internal bremsstrahlung" that arises as the emitted beta particle (electron) interacts with the charge of the proton in an electromagnetic way. In this process, some of the decay energy is carried away as photon energy. Internal bremsstrahlung gamma ray production is also a minor feature of beta decays of bound neutrons, that is, those within a nucleus.

A very small minority of neutron decays (about four per million) are so-called "two-body (neutron) decays", in which a proton, electron and antineutrino are produced as usual, but the electron fails to gain the 13.6 eV necessary energy to escape the proton (the ionization energy of hydrogen), and therefore simply remains bound to it, as a neutral hydrogen atom (one of the "two bodies"). In this type of free neutron decay, in essence all of the neutron decay energy is carried off by the antineutrino (the other "body").

The transformation of a free proton to a neutron (plus a positron and a neutrino) is energetically impossible, since a free neutron has a greater mass than a free proton.

Neutron Lifetime Puzzle

While the neutron lifetime has been studied for decades, there currently exists a lack of consilience on its exact value, due to different results from two experimental methods ("bottle" versus "beam"). While the error margin was once overlapping, increasing refinement in technique which should have resolved the issue has failed to demonstrate convergence to a single value. The difference in mean lifetime values obtained as of 2014 was approximately 9 seconds. Further, a prediction of the value based on quantum chromodynamics as of 2018 is still not sufficiently precise to support one over the other. As discussed in , the beam test would be incorrect if there is a decay mode that does not produce a proton.

Electroweak Theory

In physics, Electroweak Theory is the theory that describes both the electromagnetic force and the weak force. Superficially, these forces appear quite different. The weak force acts only across distances smaller than the atomic nucleus, while the electromagnetic force can extend for great distances (as observed in the light of stars reaching across entire galaxies), weakening only with the square of the distance. Moreover, comparison of the strength of these two fundamental interactions between two protons, for instance, reveals that the weak force is some 10 million times weaker than the electromagnetic force. Yet one of the major discoveries of the 20th century has been that these two forces are different facets of a single, more-fundamental electroweak force.

The electroweak theory arose principally out of attempts to produce a self-consistent gauge theory for the weak force, in analogy with quantum electrodynamics (QED), the successful modern theory of the electromagnetic force developed during the 1940s. There are two basic requirements for the gauge theory of the weak force. First, it should exhibit an underlying mathematical symmetry, called gauge invariance, such that the effects of the force are the same at different points in space and time. Second, the theory should be renormalizable; i.e., it should not contain nonphysical infinite quantities.

During the 1960s Sheldon Lee Glashow, Abdus Salam, and Steven Weinberg independently discovered that they could construct a gauge-invariant theory of the weak force, provided that they also included the electromagnetic force. Their theory required the existence of four massless "messenger" or carrier particles, two electrically charged and two neutral, to mediate the unified electroweak interaction. The short range of the weak force indicates, however, that it is carried by massive particles. This implies that the underlying symmetry of the theory is hidden, or "broken," by some mechanism that gives mass to the particles exchanged in weak interactions but not to the photons exchanged in electromagnetic interactions. The assumed mechanism involves an additional interaction with an otherwise unseen field, called the Higgs field, that pervades all space.

In the early 1970s Gerardus't Hooft and Martinus Veltman provided the mathematical foundation to renormalize the unified electroweak theory proposed earlier by Glashow, Salam, and Weinberg. Renormalization removed the physical inconsistencies inherent in earlier calculations of the properties of the carrier particles, permitted precise calculations of their masses, and led to more-general acceptance of the electroweak theory. The existence of the force carriers, the neutral Z particles and the charged W particles, was verified experimentally in 1983 in high-energy proton-antiproton collisions at the European Organization for Nuclear Research (CERN). The masses of the particles were consistent with their predicted values.

Beta Decay

The strong force binds particles together; by binding quarks within protons and neutrons, it indirectly binds protons and neutrons together to form nuclei. Nuclei can, however, break apart, or decay, naturally in the process known as radioactivity. One type of radioactivity, called beta decay, in which a nucleus emits an electron and thereby increases its net positive charge by one unit, has been known since the late 1890s; but it was only with the discovery of the neutron in 1932 that physicists could begin to understand correctly what happens in this radioactive process.

The most basic form of beta decay involves the transmutation of a neutron into a proton, accompanied by the emission of an electron to keep the balance of electric charge. In addition, as Wolfgang Pauli realized in 1930, the neutron emits a neutral particle that shares the energy released by the decay. This neutral particle has little or no mass and is now known to be an antineutrino, the antiparticle of the neutrino. On its own, a neutron will decay in this way after an average lifetime of 15 minutes; only within the confines of certain nuclei does the balance of forces prevent neutrons from decaying and thereby keep the entire nucleus stable.

Universal Weak Force

The rates of nuclear decay indicate that any force involved in beta decay must be much weaker than the force that binds nuclei together. It may seem counterintuitive to think of a nuclear force that can disrupt the nucleus; however, the transformation of a neutron into a proton that occurs in neutron decay is comparable to the transformations by exchange of pions that Yukawa suggested to explain the nuclear binding force. Indeed, Yukawa's theory originally tried to explain both kinds of phenomena—weak decay and strong binding—with the exchange of a single type of particle. To give the different strengths, he proposed that the exchange particle couples strongly to the heavy neutrons and protons and weakly to the light electrons and neutrinos.

Yukawa was foreshadowing future developments in unifying the two nuclear forces in this way; however, as is explained below, he had chosen the wrong two forces. He was also bold in incorporating two "new" particles in his theory—the necessary exchange particle and the neutrino predicted by Pauli only five years previously.

Pauli had been hesitant in suggesting that a second particle must be emitted in beta decay, even though that would explain why the electron could leave with a range of energies. Such was the prejudice against the prediction of new particles that theorists as eminent as Danish physicist Niels Bohr preferred to suggest that the law of conservation of energy might break down at subnuclear distances.

By 1935, however, Pauli's new particle had found a champion in Enrico Fermi. Fermi named the particle the neutrino and incorporated it into his theory for beta decay, published in 1934. Like Yukawa, Fermi drew on an analogy with QED; but Fermi regarded the emission of the neutrino and electron by the neutron as the direct analog of the emission of a photon by a charged particle, and he did not invoke a new exchange particle. Only later did it become clear that, strictly speaking, the neutron emits an antineutrino.

Fermi's theory, rather than Yukawa's, proved highly successful in describing nuclear beta decay, and it received added support in the late 1940s with the discovery of the pion and its relationship with the muon. In particular, the muon decays to an electron, a neutrino, and an antineutrino in a process that has exactly the same basic strength as the neutron's decay to a proton. The idea of a "universal" weak interaction that, unlike the strong force, acts equally upon light and heavy particles (or leptons and hadrons) was born.

Early Theories

The nature of the weak force began to be further revealed in 1956 as the result of work by two Chinese American theorists, Tsung-Dao Lee and Chen Ning Yang. Lee and Yang were trying to resolve some puzzles in the decays of the strange particles. They discovered that they could solve the mystery, provided that the weak force does not respect the symmetry known as parity.

The parity operation is like reflecting something in a mirror; it involves changing the coordinates (x, y, z) of each point to the "mirror" coordinates (−x, −y, −z). Physicists had always assumed that such an operation would make no difference to the laws of physics. Lee and Yang, however, proposed that the weak force is exceptional in this respect, and they suggested ways that parity violation might be observed in weak interactions. Early in 1957, just a few months after Lee and Yang's theory was published, experiments involving the decays of neutrons, pions, and muons showed that the weak force does indeed violate parity symmetry. Later that year Lee and Yang were awarded the Nobel Prize for Physics for their work.

Parity violation and the concept of a universal form of weak interaction were combined into one theory in 1958 by the American physicists Murray Gell-Mann and Richard Feynman. They established the mathematical structure of the weak interaction in what is known as V−A, or vector minus axial vector, theory. This theory proved highly successful experimentally, at least at the relatively low energies accessible to particle physicists in the 1960s. It was clear that the theory had the correct kind of mathematical structure to account for parity violation and related effects, but there were strong indications that, in describing particle interactions at higher energies than experiments could at the time access, the theory began to go badly wrong.

The problems with V−A theory were related to a basic requirement of quantum field theory—the existence of a gauge boson, or messenger particle, to carry the force. Yukawa had attempted to describe the weak force in terms of the same intermediary that is responsible for the nuclear binding force, but this approach did not work. A few years after Yukawa published his theory, a Swedish theorist, Oskar Klein, proposed a slightly different kind of carrier for the weak force.

In contrast to Yukawa's particle, which had spin 0, Klein's intermediary had spin 1 and therefore would give the correct spins for the antineutrino and the electron emitted in the beta decay of the neutron. Moreover, within the framework of Klein's concept, the known strength of the weak force in beta decay showed that the mass of the particle must be approximately 100 times the proton's mass, although the theory could not predict this value. All attempts to introduce such a particle into V–A theory, however, encountered severe difficulties, similar to those that had beset QED during the 1930s and early' 40s. The theory gave infinite probabilities to various interactions, and it defied the renormalization process that had been the salvation of QED.

Hidden Symmetry

Throughout the 1950s, theorists tried to construct field theories for the nuclear forces that would exhibit the same kind of gauge symmetry inherent in James Clerk Maxwell's theory of electrodynamics and in QED. There were two major problems, which were in fact related. One concerned the infinities and the difficulty in renormalizing these theories; the other concerned the mass of the intermediaries. Straightforward gauge theory requires particles of zero mass as carriers, such as the photon of QED, but Klein had shown that the short-ranged weak force requires massive carriers.

Physicists had to discover the correct mathematical symmetry group for describing the transformations between different subatomic particles and then identify for the known forces the messenger particles required by fields with the chosen symmetry. Early in the 1960s Sheldon Glashow in the United States and Abdus Salam and John Ward in England decided to work with a combination of two symmetry groups—namely, SU(2) × U(1). Such a symmetry requires four spin-1 messenger particles, two electrically neutral and two charged. One of the neutral particles could be identified with the photon, while the two charged particles could be the messengers responsible for beta decay, in which charge changes hands, as when the neutron decays into a proton. The fourth messenger, a second neutral particle, seemed at the time to have no obvious role; it apparently would permit weak interactions with no change of charge—so-called neutral current interactions—which had not yet been observed.

However, still required the messengers to be massless, which was all right for the photon but not for the messengers of the weak force. Toward the end of the 1960s, Salam and Steven Weinberg, an American theorist, independently realized how to introduce massive messenger particles into the theory while at the same time preserving its basic gauge symmetry properties. The answer lay in the work of the English theorist Peter Higgs and others, who had discovered the concept of symmetry breaking, or, more descriptively, hidden symmetry.

A physical field can be intrinsically symmetrical, although this may not be apparent in the state of the universe in which experiments are conducted. On Earth's surface, for example, gravity seems asymmetrical—it always pulls down. From a distance, however, the symmetry of the gravitational field around Earth becomes apparent. At a more-fundamental level, the fields associated with the electromagnetic and weak forces are not overtly symmetrical, as is demonstrated by the widely differing strengths of weak and electromagnetic interactions at low energies. Yet, according to

Higgs's ideas, these forces can have an underlying symmetry. It is as if the universe lies at the bottom of a wine bottle; the symmetry of the bottle's base is clear from the top of the dimple in the centre, but it is hidden from any point in the valley surrounding the central dimple.

Higgs's mechanism for symmetry breaking provided Salam and Weinberg with a means of explaining the masses of the carriers of the weak force. Their theory, however, also predicted the existence of one or more new "Higgs" bosons, which would carry additional fields needed for the symmetry breaking and would have spin 0. With this sole proviso the future of the electroweak theory began to look more promising. In 1971 a young Dutch theorist, Gerardus 't Hooft, building on work by Martinus Veltman, proved that the theory is renormalizable (in other words, that all the infinities cancel out). Many particle physicists became convinced that the electroweak theory was, at last, an acceptable theory for the weak force.

Finding the Messenger Particles

In addition to the Higgs boson, or bosons, electroweak theory also predicts the existence of an electrically neutral carrier for the weak force. This neutral carrier, called the Z^0, should mediate the neutral current interactions—weak interactions in which electric charge is not transferred between particles. The search for evidence of such reactions, which would confirm the validity of the electroweak theory, began in earnest in the early 1970s.

The first signs of neutral currents came in 1973 from experiments at the European Organization for Nuclear Research (CERN) near Geneva. A team of more than 50 physicists from a variety of countries had diligently searched through the photographs taken of tracks produced when a large bubble chamber called Gargamelle was exposed to a beam of muon-antineutrinos. In a neutral current reaction an antineutrino would simply scatter from an electron in the liquid contents of the bubble chamber. The incoming antineutrino, being neutral, would leave no track, nor would it leave a track as it left the chamber after being scattered off an electron. But the effect of the neutral current—the passage of a virtual Z^0 between the antineutrino and the electron—would set the electron in motion, and, being electrically charged, the electron would leave a track, which would appear as if from nowhere. Examining approximately 1.4 million pictures, the researchers found three examples of such a neutral current reaction. Although the reactions occurred only rarely, there were enough to set hopes high for the validity of electroweak theory.

In 1979 Glashow, Salam, and Weinberg, the theorists who had done much of the work in developing electroweak theory in the 1960s, were awarded the Nobel Prize for Physics; 't Hooft and Veltman were similarly rewarded in 1999. By that time, enough information on charged and neutral current interactions had been compiled to predict that the masses of the weak messengers required by electroweak theory should be about 80 gigaelectron volts (GeV; 109 eV) for the charged W+ and W− particles and 90 GeV for the Z^0. There was, however, still no sign of the direct production of the weak messengers, because no accelerator was yet capable of producing collisions energetic enough to create real particles of such large masses (nearly 100 times as massive as the proton).

A scheme to find the W and Z particles was under way at CERN, however. The plan was to accelerate protons in one direction around CERN's largest proton synchrotron (a circular accelerator) and antiprotons in the opposite direction. At an appropriate energy (initially 270 GeV per beam), the two sets of particles would be made to collide head-on. The total energy of the collision would be far greater than anything that could be achieved by directing a single beam at a stationary target, and physicists hoped it would be sufficient to produce a small but significant number of W and Z particles.

In 1983 the researchers at CERN, working on two experiments code-named UA1 and UA2, were rewarded with the discovery of the particles they sought. The Ws and Zs that were produced did not live long enough to leave tracks in the detectors, but they decayed to particles that did leave tracks. The total energy of those decay particles, moreover, equaled the energy corresponding to the masses of the transient W and Z particles, just as predicted by electroweak theory. It was a triumph both for CERN and for electroweak theory. Hundreds of physicists and engineers were involved in the project, and in 1984 the Italian physicist Carlo Rubbia and Dutch engineer Simon van der Meer received the Nobel Prize for Physics for their leading roles in making the discovery of the W and Z particles possible.

The W particles play a crucial role in interactions that turn one flavour of quark or lepton into another, as in the beta decay of a neutron, where a down quark turns into an up quark to form a proton. Such flavour-changing interactions occur only through the weak force and are described by the SU(2) symmetry that underlies electroweak theory along with U(1). The basic representation of this mathematical group is a pair, or doublet, and, according to electroweak theory, the quarks and leptons are each grouped into pairs of increasing mass: (u, d), (c, s), (t, b) and (e, ve), (μ, vμ), (τ, vτ). This underlying symmetry does not, however, indicate how many pairs of quarks and leptons should exist in total. This question was answered in experiments at CERN in 1989, when the colliding-beam storage ring particle accelerator known as the Large Electron-Positron (LEP) collider came into operation.

When LEP started up, it could collide electrons and positrons at total energies of about 90 GeV, producing copious numbers of Z particles. Through accurate measurements of the "width" of the Z—that is, the intrinsic variation in its mass, which is related to the number of ways the particle can decay—researchers at the LEP collider have found that the Z can decay to no more than three types of light neutrino. This in turn implies that there are probably no more than three pairs of leptons and three pairs of quarks.

Weak Isospin

In particle physics, weak isospin is a quantum number relating to the weak interaction, and parallels the idea of isospin under the strong interaction. Weak isospin is usually given the symbol T or I with the third component written as $T_z, T_3, I_z or\ I_3$. It can be understood as the eigenvalue of a charge operator.

The weak isospin conservation law relates the conservation of T_3; all weak interactions must preserve T_3. It is also conserved by the electromagnetic, and strong interactions. However, one of the interactions is with the Higgs field. Since the Higgs field vacuum expectation value is nonzero, particles interact with this field all the time even in vacuum. This changes their weak isospin (and weak hypercharge). Only a specific combination of them, $Q = T_3 + \frac{1}{2}Y_W$ (electric charge), is conserved. T_3 is more important than T and often the term "weak isospin" refers to the "3rd component of weak isospin".

Relation with Chirality

Fermions with negative chirality (also called "left-handed" fermions) have $T_3 = +\frac{1}{2}$ and can be grouped into doublets with $T_3 = +\frac{1}{2}$ that behave the same way under the weak interaction. For example, up-type quarks (u, c, t) have $T_3 = +\frac{1}{2}$ and always transform into down-type quarks (d, s, b), which have $T_3 = -\frac{1}{2}$, and vice versa. On the other hand, a quark never decays weakly into a quark of the same T_3. Something similar happens with left-handed leptons, which exist as doublets containing a charged lepton (e^-, μ^-, τ^-) with $T_3 = \pm\frac{1}{2}$ and a neutrino $(\nu_e, \nu_\mu, \nu_\tau)$ with $T_3 = +\frac{1}{2}$. In all cases, the corresponding *anti*-fermion has reversed chirality ("right-handed" antifermion) and sign reversed T_3.

Fermions with positive chirality ("right-handed" fermions) and *anti*-fermions with negative chirality ("left-handed" anti-fermions) have $T = T_3 = 0$ and form singlets that *do not undergo weak interactions.*

The electric charge, Q, is related to weak isospin, T_3, and weak hypercharge, by Y_W $Q = T_3 + \frac{1}{2}Y_W$.

Left-handed fermions in the Standard Model								
Generation 1			Generation 2			Generation 3		
Fermion	Symbol	Weak isospin	Fermion	Symbol	Weak isospin	Fermion	Symbol	Weak isospin
Electron neutrino	ν_e	$+\frac{1}{2}$	Muon neutrino	ν_μ	$+\frac{1}{2}$	Tau neutrino	ν_τ	$+\frac{1}{2}$
Electron	e^-	$-\frac{1}{2}$	Muon	μ^-	$-\frac{1}{2}$	Tau	τ^-	$-\frac{1}{2}$
Up quark	u	$+\frac{1}{2}$	Charm quark	c	$+\frac{1}{2}$	Top quark	t	$+\frac{1}{2}$
Down quark	d	$-\frac{1}{2}$	Strange quark	s	$-\frac{1}{2}$	Bottom quark	b	$-\frac{1}{2}$
All of the above left-handed (*regular*) particles have corresponding right-handed *anti*-particles with equal and opposite weak isospin.								
All right-handed (regular) particles and left-handed antiparticles have weak isospin of 0.								

Weak Isospin and the W Bosons

The symmetry associated with weak isospin is SU(2) and requires gauge bosons with $T = 1(W^+ W^- and W^0)$ to mediate transformations between fermions with half-integer weak isospin charges. $T = 1$ implies that W bosons have three different values of T_3:

- W^+ boson $(T_3 = +1)$ is emitted in transitions $(T_3 = +\frac{1}{2}) \to (T_3 = -\frac{1}{2})$
- W^0 boson $(T_3 = 0)$ would be emitted in weak interactions where T_3 does not change, such as neutrino scattering.

- W⁻ boson ($T_3 = -1$) is emitted in transitions $(T_3 = -\tfrac{1}{2}) \rightarrow (T_3 = +\tfrac{1}{2})$

Under electroweak unification, the W^0 boson mixes with the weak hypercharge gauge boson B, resulting in the observed Z^0boson and the photon of quantum electrodynamics; the resulting $Z^0$and the photon both have weak isospin = 0.

The sum of −isospin and +charge is zero for each of the bosons, consequently, all the electroweak bosons have weak hypercharge $Y_W = 0$, so unlike gluons of the color force, the electroweak bosons are unaffected by the force they mediate.

Weak Hypercharge

In the Standard Model of electroweak interactions of particle physics, the weak hypercharge is a quantum number relating the electric charge and the third component of weak isospin. It is frequently denoted Y_W and corresponds to the gauge symmetry U(1).

It is conserved (only terms that are overall weak-hypercharge neutral are allowed in the Lagrangian). However, one of the interactions is with the Higgs field. Since the Higgs field vacuum expectation value is nonzero, particles interact with this field all the time even in vacuum. This changes their weak hypercharge (and weak isospin T_3). Only a specific combination of them, $Q = T_3 + 1/2Y_W$ (electric charge), is conserved.

Mathematically, weak hypercharge appears similar to the Gell-Mann–Nishijima formula for the hypercharge of strong interactions (which is not conserved in weak interactions) and which does not apply to leptons.

Weak hypercharge is the generator of the U(1) component of the electroweak gauge group, $SU(2)\times U(1)$ and its associated quantum field B mixes with the W^3 electroweak quantum field to produce the observed Z gauge boson and the photon of quantum electrodynamics.

The Weak hypercharge satisfies the relation

$$Q = T_3 + \tfrac{1}{2}Y_W ,$$

where Q is the electric charge (in elementary charge units) and T_3 is the third component of weak isospin (the SU(2) component).

Rearranging, the weak hypercharge can be explicitly defined as:

$$Y_W = 2(Q - T_3)$$

Fermion family	Left-chiral fermions				Right-chiral fermions			
		Electric charge Q	Weak isospin T_3	Weak hypercharge Y_W		Electric charge Q	Weak isospin T_3	Weak hypercharge Y_W

Leptons	ν_e, ν_μ, ν_τ	0	$+\frac{1}{2}$	−1	No interaction, if extant			
	e^-, μ^-, τ^-	−1	$-\frac{1}{2}$	−1	e^-_R, μ^-_R, τ^-_R	−1	0	−2
Quarks	u, c, t	$+\frac{2}{3}$	$+\frac{1}{2}$	$+\frac{1}{3}$	u_R, c_R, t_R	$+\frac{2}{3}$	0	$+\frac{4}{3}$
	d, s, b	$-\frac{1}{3}$	$-\frac{1}{2}$	$+\frac{1}{3}$	d_R, s_R, b_R	$-\frac{1}{3}$	0	$-\frac{2}{3}$

where "left"- and "right"-handed here are left and right chirality, respectively (distinct from helicity).

Mediated fundamental interaction	**Boson**	**Electric charge Q**	**Weak isospin T_3**	**Weak hypercharge Y_W**
Weak	W	±1	±1	0
	Z	0	0	0
Electric	γ	0	0	0
Higgs	H^0	0	$-\frac{1}{2}$	+1

Hypercharge assignments in the Standard Model are determined up to a twofold ambiguity by requiring cancellation of all anomalies.

Alternative Scale

For convenience, weak hypercharge is often represented at half-scale, so that

$$Y_W = Q - T_3,$$

which is equal to just *the average electric charge of the particles in the isospin multiplet.*

Baryon and Lepton Number

Weak hypercharge is related to baryon number minus lepton number via:

$$\tfrac{1}{2}X + Y_W = \tfrac{5}{2}(B - L)$$

where X is a conserved quantum number in GUT. Since weak hypercharge is always conserved this implies that baryon number minus lepton number is also always conserved, within the Standard Model and most extensions.

Neutron Decay

$$n \rightarrow p + e^- + \overline{\nu}_e$$

Hence neutron decay conserves baryon number B and lepton number L separately, so also the difference $B - L$ is conserved.

Proton Decay

Proton decay is a prediction of many grand unification theories.

$$p^+ \rightarrow e^+ + \pi^0 \rightarrow e^+ + 2\gamma$$

Hence proton decay conserves $B - L$, even though it violates both lepton number and baryon number conservation.

References

- Electroweak-theory, science: britannica.com, Retrieved 16 May, 2019
- Basic Ideas and Concepts in Nuclear Physics: An Introductory Approach, Third Edition K. Heyde Taylor & Francis 2004. Print ISBN 978-0-7503-0980-6. eBook ISBN 978-1-4200-5494-1. DOI: 10.1201/9781420054941.ch5
- Quantum-chromodynamics-Describing-the-strong-force, science-subatomic-particle- 60750: britannica.com, Retrieved 06 February, 2019
- Greene, Geoffrey L.; Geltenbort, Peter (2016). "The Neutron Enigma". Scientific American. 314 (4): 36–41. Bibcode:2016SciAm.314d..36G. doi:10.1038/scientificamerican0416-36. ISSN 0036-8733
- T. P. Cheng; L. F. Li (2006). Gauge theory of elementary particle physics. Oxford University Press. ISBN 0-19-851961-3

PERMISSIONS

All chapters in this book are published with permission under the Creative Commons Attribution Share Alike License or equivalent. Every chapter published in this book has been scrutinized by our experts. Their significance has been extensively debated. The topics covered herein carry significant information for a comprehensive understanding. They may even be implemented as practical applications or may be referred to as a beginning point for further studies.

We would like to thank the editorial team for lending their expertise to make the book truly unique. They have played a crucial role in the development of this book. Without their invaluable contributions this book wouldn't have been possible. They have made vital efforts to compile up to date information on the varied aspects of this subject to make this book a valuable addition to the collection of many professionals and students.

This book was conceptualized with the vision of imparting up-to-date and integrated information in this field. To ensure the same, a matchless editorial board was set up. Every individual on the board went through rigorous rounds of assessment to prove their worth. After which they invested a large part of their time researching and compiling the most relevant data for our readers.

The editorial board has been involved in producing this book since its inception. They have spent rigorous hours researching and exploring the diverse topics which have resulted in the successful publishing of this book. They have passed on their knowledge of decades through this book. To expedite this challenging task, the publisher supported the team at every step. A small team of assistant editors was also appointed to further simplify the editing procedure and attain best results for the readers.

Apart from the editorial board, the designing team has also invested a significant amount of their time in understanding the subject and creating the most relevant covers. They scrutinized every image to scout for the most suitable representation of the subject and create an appropriate cover for the book.

The publishing team has been an ardent support to the editorial, designing and production team. Their endless efforts to recruit the best for this project, has resulted in the accomplishment of this book. They are a veteran in the field of academics and their pool of knowledge is as vast as their experience in printing. Their expertise and guidance has proved useful at every step. Their uncompromising quality standards have made this book an exceptional effort. Their encouragement from time to time has been an inspiration for everyone.

The publisher and the editorial board hope that this book will prove to be a valuable piece of knowledge for students, practitioners and scholars across the globe.

INDEX

www.ingramcontent.com/pod-product-compliance
Lightning Source LLC
Chambersburg PA
CBHW082043190326
41458CB00010B/3446
* 9 7 8 1 6 4 1 7 2 6 8 5 6 *